Fundamentals of Motion: Introduction to Classical Mechanics

FUNDAMENTALS OF MOTION: INTRODUCTION TO CLASSICAL MECHANICS

Prof. Ravi Prakash

ARCLER
PRESS

www.arclerpress.com

Fundamentals of Motion: Introduction to Classical Mechanics

Prof. Ravi Prakash

Arcler Press
224 Shoreacres Road
Burlington, ON L7L 2H2
Canada
www.arclerpress.com
Email: orders@arclereducation.com

ISBN: 978-1-77469-937-9 (Hardcover)

ABOUT THE AUTHOR

Prof. Ravi Prakash is working as an assisastant Professor in Electronics and Communication Engineering Uma Nath Singh Institute of Engineering and Technology Veer Bahadur Singh Purvanchal University Jaunpur, Uttar Pradesh, India. He was born on January 19, 1975, in Jaunpur, U.P., India. He received the B. Tech. degree in Electronics and Telecommunication Engineering from University of Allahabad, Allahabad, U.P., India in 1997, and M. Tech. degree in Electronics Engineering (Communication Technology) from University of Allahabad, Allahabad, U. P., 2001. He is gold medalist at his graduation level and rank holder in board examination. He has almost 21 years teaching experience at B. Tech., M. Tech. and Ph.D. level in engineering. He had worked as head of department of Electronics Engineering at Veer Bahadur Singh Purvanchal University Jaunpur for almost ten years. He received the Ph.D. degree in Electronics and Communication Engineering from University of Allahabad, U.P., India. His current research interests include Communication Technology. He has published several research articles in reputed international journals related to wireless, optical and digital communication. He has worked as several research and reviewer committee member, and he has excellent work in the field of digital communication, wireless and mobile communication. He had worked as Nodal officer, academic in the Technical Education Quality Improvement Program (TEQIP) supported by World Bank at university level.

TABLE OF CONTENTS

LIST OF FIGURES

LIST OF TABLES

PREFACE

The Fundamentals of Motion in Classical Mechanics refer to the principles that describe the motion of objects in our everyday world. Rooted in Sir Isaac Newton's laws of motion, Classical Mechanics provides a framework for understanding how forces act on physical bodies and how those bodies move as a result. The first law, known as the law of inertia, states that an object will remain at rest or in uniform motion unless acted upon by an external force. The second law establishes a connection between force, mass, and acceleration, providing the famous equation F=ma. The third law, the principle of action and reaction, holds that for every action, there is an equal and opposite reaction. Together, these laws enable precise predictions of motion for macroscopic objects at speeds much slower than the speed of light. Classical Mechanics fails at extremely small scales (where Quantum Mechanics is needed) or near the speed of light (where Relativity becomes essential), but for everyday situations, it remains a vital and accurate model.

"Fundamentals of Motion: Introduction to Classical Mechanics" is an insightful and comprehensive guide designed to provide readers with an in-depth understanding of the fundamental principles and phenomena that govern the motion of objects. The book is structured into eight carefully crafted chapters, each focusing on a distinct aspect of classical mechanics.

Chapter 1 serves as an introduction to classical mechanics, laying the groundwork for what is to come by presenting fundamental concepts and historical context (2 units). Chapter 2, titled "Kinematics: Describing Motion," dedicates itself to the science of motion without regard to forces (1 unit).

In Chapter 3, "Dynamics: Forces and Motion," the focus shifts to the relationship between forces and the motion they induce, building a bridge between kinematics and dynamics. Chapter 4 discusses the key concepts of "Work, Energy, and Power," elaborating on their interconnectedness.

A significant portion of the book, Chapter 5, delves into "Momentum and Impulse and Collision," providing an extensive examination of these fundamental principles and their applications in various fields. Chapter 6, focusing on "Rotational Motion," discusses the motion of objects around a fixed axis.

Chapter 7 introduces "Oscillations and Simple Harmonic Motion," exploring periodic motions and the mathematical models that describe them. Finally, Chapter 8 encapsulates the principles of "Universal Gravitation and Planetary Motion," elucidating the laws that govern the motion of celestial bodies.

"Fundamentals of Motion: Introduction to Classical Mechanics" represents an invaluable resource for students, educators, and professionals alike, offering a meticulous and methodical exposition of classical mechanics. By traversing the realm of motion from basic concepts to advanced theories, this book aims to cultivate a profound understanding and appreciation for the underlying mechanics that govern our physical world. Its precise structure, expertly authored content, and thoughtful pedagogy make it an indispensable tool for those seeking to master the principles and applications of classical mechanics.

CHAPTER 1

Introduction to Classical Mechanics

Contents

1.1. INTRODUCTION

Mechanics, as a field within the realm of physics, focuses on examining motion, forces, and the responses exhibited by physical entities when subjected to diverse external and internal stimuli. The subject matter is the fundamental basis of classical physics and holds significant importance in comprehending the fundamental principles governing objects' motion within the universe. The discipline of mechanics encompasses entities at various scales, ranging from macroscopic objects like planets and vehicles to subatomic particles. Fundamentally, the field of mechanics aims to address inquiries about the reasons and mechanisms behind the movement of objects, the variables that influence their motion, and the mathematical principles that elucidate their actions. Through the development of mathematical models and principles, the field of mechanics enables the prediction and analysis of object motion, rendering it an essential instrument for professionals in engineering, physics, and various scientific domains (Cunningham, 2013).

The fundamental aim of mechanics is to provide an accurate description and explanation of motion, thereby serving as an essential foundation for comprehending more intricate phenomena within physics. By engaging in the systematic practice of observing and analyzing the motion of objects, the discipline of mechanics provides valuable knowledge regarding the fundamental principles that govern the behavior of the physical world. Classical mechanics primarily focuses on scenarios characterized by non-relativistic conditions, where the velocities of objects are considerably smaller than the speed of light. Mechanics is the cornerstone of classical physics, providing the necessary tools to understand the motion of objects and the forces that influence them. By defining the fundamental principles of motion, such as Newton's laws of motion, mechanics offers a comprehensive framework for studying the physical behavior of the world around us. Through rigorous observation, experimentation, and mathematical analysis, mechanics continues to shape our understanding of motion and the fundamental laws governing the universe (Taylor & Taylor, 2005).

1.2. IMPORTANCE OF STUDYING CLASSICAL MECHANICS

Studying classical mechanics is paramount in physics and engineering due to its foundational role in understanding the fundamental principles that govern motion and forces. Here are some key reasons why studying classical mechanics is crucial:

1. Predicting and Analyzing Motion: Classical mechanics offers a systematic and precise approach to predicting the motion of objects under the influence of forces. By understanding the mathematical relationships between position, velocity, acceleration, and time, we can accurately predict the future motion of objects. This predictive power is invaluable in designing and analyzing various systems, from simple mechanical devices to complex astronomical phenomena (Lamanepa et al., 2022).
2. Foundation for Advanced Physics: Classical mechanics is a foundation for more advanced branches of physics, such as quantum mechanics and relativity. While classical mechanics is limited to describing non-relativistic situations, it provides essential insights into the behavior of matter and energy, laying the groundwork for exploring the complexities of the subatomic and cosmic realms.
3. Engineering Applications: Engineers extensively use classical mechanics to design and optimize structures, machines, and systems. Understanding the forces and motion involved in various engineering projects helps ensure safety, efficiency, and reliability. Classical mechanics principles are essential in civil engineering, aerospace engineering, mechanical engineering, and many others (Henneaux & Teitelboim, 1988).
4. Problem-Solving Skills: Studying classical mechanics sharpens problem-solving skills. It challenges students to apply mathematical concepts and principles to real-world scenarios, fostering critical thinking and analytical abilities. Analyzing complex systems and breaking them into manageable components is transferable to many other fields.
5. Historical Significance: Classical mechanics has a rich history, with contributions from eminent scientists like Galileo, Newton, and Euler. Understanding the development of classical mechanics and its historical context provides a deeper appreciation of the scientific method and the progress of human knowledge (Kolmogorov & Tichomirov, 1991).
6. Practical Applications: Classical mechanics principles are evident in everyday experiences, such as the motion of vehicles, the flight of projectiles, and the behavior of sports equipment. A grasp of classical mechanics enhances our understanding and appreciation of the world.

Studying classical mechanics is essential for grasping the fundamental principles of motion and the forces that govern the physical world. Its predictive power, interdisciplinary applications, problem-solving benefits, and historical significance make it an indispensable subject for anyone seeking to comprehend the underlying mechanics of nature and technology (Berry, 1977).

1.3. HISTORICAL OVERVIEW OF CLASSICAL MECHANICS

Classical mechanics has a rich history that spans over two millennia, with significant contributions from prominent thinkers and scientists. From the ancient Greek philosophers to the groundbreaking work of Isaac Newton, the development of classical mechanics revolutionized our understanding of motion and laid the groundwork for modern physics. This historical overview highlights the key contributions of Aristotle, Galileo, and Newton to classical mechanics (Di Mauro & Naddeo, 2022).

1.3.1. Aristotle's Theory of Motion

In ancient Greece, around the 4th century BCE, Aristotle proposed a comprehensive theory of motion based on his observations and philosophical reasoning. Aristotle's theory centered on the concept of natural motion and violent motion. According to Aristotle, all objects in the universe had a "natural place" to which they sought to move. For example, earth and water were believed to have a natural tendency to move downward, while fire and air were thought to move upward. This theory of natural motion formed the basis of Aristotle's view on celestial mechanics as well, where he proposed that the heavens were composed of a fifth element called "aether" and that celestial bodies moved in perfect circular orbits (Figure 1.1) (Balaban, 1995).

Furthermore, Aristotle is credited with introducing the concept of "violent motion," which elucidates the phenomenon of motion that arises due to external forces exerted upon objects. The individual posited that to maintain motion, an object necessitated an unbroken external force, referred to as "the mover." Although Aristotle's theory held significant sway in his era, it harbored certain inaccuracies, and its geocentric perspective persisted until the advent of the Renaissance (Graham, 1988).

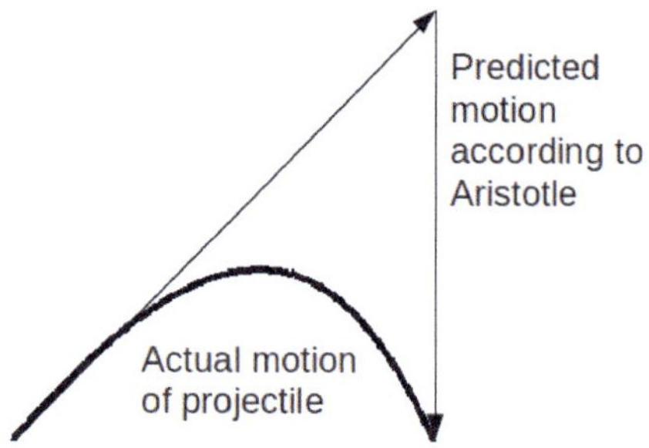

Figure 1.1. Aristotle's conception of motion (Source: A Star Maths and Physics, Creative Commons License)

1.3.2. Galileo's Contributions to Mechanics

During the 16th and 17th centuries, the renowned Italian scientist Galileo Galilei made significant and pioneering advancements in mechanics. He engaged in methodical experimentation and meticulous observation, frequently challenging the prevailing Aristotelian perspectives. One of the fundamental contributions made by Galileo was his investigation and analysis of the concept of inertia. The individual postulated that objects in motion would persist in their motion unless subjected to an external force. In contrast, objects at rest would remain stationary unless a force was applied to initiate their motion.

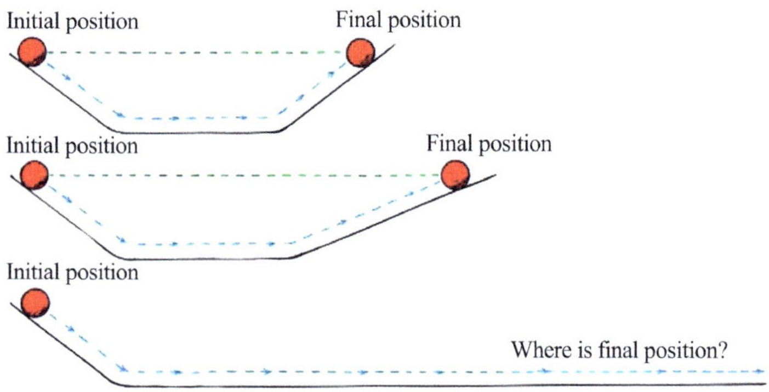

Figure 1.2. Galileo's experiment of motion (Source: Emma Garrett, Creative Commons License)

The proposition presented herein directly challenges Aristotle's theory regarding natural motion. The experiments conducted by Galileo involving inclined planes provided evidence that the acceleration of falling objects is not influenced by their mass, which contradicted the prevailing Aristotelian viewpoint. Additionally, he made notable progress in projectile motion by demonstrating the independence of horizontal and vertical motion components. The insights above served as the basis for formulating and establishing the laws of motion, attributed to the renowned scholar Sir Isaac Newton (Figure 1.2) (Wallace, 1971).

1.3.3. Newton's Laws of Motion

In the late 17th century, Sir Isaac Newton formulated the three fundamental laws of motion, which revolutionized the study of mechanics and became the cornerstone of classical physics.

a. According to Newton's first law of motion, also known as the law of inertia, it is postulated that an object initially at rest will continue to remain in that state. In contrast, an object in motion will continue to move in a straight line at a constant velocity unless it is subjected to the influence of an external net force. The passage of this legislation effectively rendered Aristotle's concept of "natural motion" outdated while simultaneously introducing the fundamental principle of inertia as a viable alternative. This principle is the building block for understanding objects' behavior in the space vacuum when no external forces exist (Nowakowski & Rosu, 2002).

b. The second law of motion, established by Isaac Newton, outlines a basic link between force, mass, and acceleration. According to the hypothesis, the total force applied to an object is proportional to its mass and the acceleration it experiences due to that force. This rule is mathematically represented by the equation $F = ma$, in which F stands for the total force acting on the object, m stands for the object's mass, and the acceleration the object feels (White, 1984).

c. Newton's Third Law of Motion (Law of Action and Reaction): Newton's third law states that for every action, there is an equal and opposite reaction. When one object exerts a force on another object, the second object exerts an equal force in the opposite direction on the first object. This law highlights the interconnectedness of forces in nature and is fundamental in explaining the behavior of interacting systems (Saglam-Arslan & Devecioglu, 2010).

Remember

Newton's laws of motion are fundamental principles in classical mechanics, and they provide a powerful framework for understanding the relationships between forces and motion.

The laws of motion formulated by Sir Isaac Newton offer a comprehensive and precise theoretical framework for comprehending the dynamics of objects, encompassing their motion on Earth and in celestial bodies. The contributions of his work established the fundamental principles of classical mechanics. They functioned as a cohesive factor in examining physics and engineering, stimulating subsequent advancements in science and technology (Figure 1.3) (Waldrip et al., 2013).

The historical examination of classical mechanics elucidates the transition from ancient philosophical concepts to methodical experimentation and mathematical frameworks, ultimately culminating in the groundbreaking contributions of Galileo and Newton. The seminal contributions made by these pioneers have established a solid basis for contemporary physics and engineering, facilitating our understanding of the fundamental principles that govern motion and forces in the cosmos. The ongoing progression of classical mechanics has significantly influenced our comprehension of the natural world and remains an essential component of contemporary scientific inquiry (Carling et al., 1998).

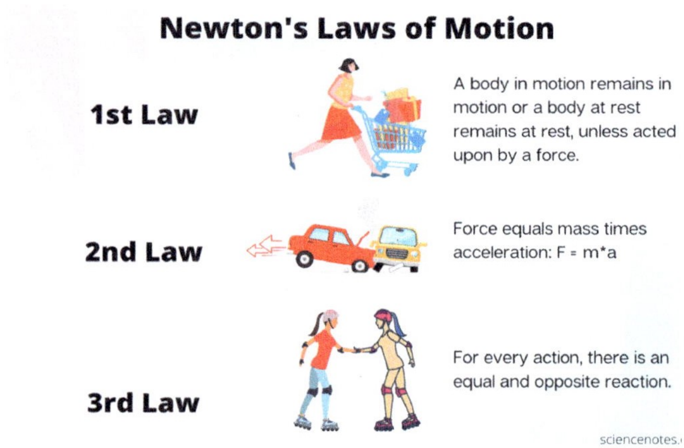

Figure 1.3. Newton's laws of motion (Source: Science Notes, Creative Commons License)

1.4. SCOPE AND APPLICATIONS OF CLASSICAL MECHANICS

Classical mechanics is a foundational branch of physics with a broad scope and numerous practical applications in various fields. It provides a comprehensive framework for understanding the behavior of macroscopic objects under the influence of forces and motion. The scope and applications of classical mechanics are vast and impactful, touching upon diverse science, engineering, and technology areas. Here are some of the key aspects of its scope and applications (Sanjuán, 2021):

1. Dynamics of Everyday Objects: Classical mechanics explains the motion of everyday objects, from the simple falling of an apple to the motion of vehicles on roads. By applying Newton's laws of motion and concepts like friction, classical mechanics helps us understand and predict the behavior of objects in our daily lives.
2. Celestial Mechanics: Classical mechanics played a significant role in the development of celestial mechanics, which focuses on the motion of celestial bodies, such as planets, moons, and comets. The application of gravitational forces, as formulated by Newton, enabled the accurate prediction of planetary orbits and celestial phenomena (Muckerman, 1981).
3. Engineering and Design: Classical mechanics is an essential tool in engineering and design. It analyzes and optimizes buildings, bridges, vehicles, and machinery structures and mechanisms. Engineers rely on classical mechanics to ensure the safety and functionality of various engineering projects.
4. Aerospace and Astronautics: The principles of classical mechanics are critical in aerospace engineering and astronautics. Understanding motion and forces is crucial for designing spacecraft, launching vehicles, and planning space missions. Calculations of trajectories and orbital mechanics heavily rely on classical mechanics principles (Vedral, 2010).
5. Fluid Mechanics: Classical mechanics also extends to fluid mechanics, which deals with the motion of fluids (liquids and gases) and their interactions with solid boundaries. Fluid mechanics is crucial in understanding phenomena like fluid flow, turbulence, and aerodynamics, all of which have practical applications in engineering and environmental sciences.

6. Sports and Biomechanics: Classical mechanics find applications in sports and biomechanics to analyze the motion of athletes and sports equipment. By understanding forces and motions in sports activities, coaches and athletes can improve performance and reduce the risk of injuries.
7. Machine Design: In machine design, classical mechanics plays a vital role in designing efficient mechanisms, analyzing forces in gears and linkages, and optimizing the operation of mechanical systems (Czarnocka & Żytkow, 1982).
8. Material Science: Classical mechanics principles are used in material science to understand the mechanical properties of materials, such as strength, elasticity, and fatigue. This knowledge is crucial in developing new materials and improving existing ones for various applications.
9. Earthquake Engineering: Classical mechanics is utilized in earthquake engineering to analyze the behavior of structures during seismic events, ensuring their resilience and safety.
10. Robotics and Automation: Classical mechanics is integral to studying robotics and automation. It allows engineers to design robotic systems with precise control of motion and forces, enabling a wide range of applications from manufacturing to space exploration (Sarlet & Cantrijn, 1981).

1.5. KINEMATICS: DESCRIBING MOTION

Kinematics is a fundamental subfield within classical mechanics primarily concerned with characterizing and examining object motion while disregarding the underlying forces responsible for inducing such motion. This subject pertains to fundamental principles encompassing position, velocity, acceleration, and time, facilitating a comprehensive comprehension of the dynamics of objects in both spatial and temporal dimensions. Kinematics employs mathematical equations and graphical representations to provide a quantitative depiction of diverse facets of motion, encompassing both straightforward unidirectional movement and intricate multidimensional motions. Kinematics allows for accurately describing an object's trajectory, velocity, and acceleration, serving as a foundational framework for studying classical mechanics and its practical implementations in diverse domains such as engineering, physics, and animation (Saltiel & Malgrange, 1980).

1.5.1. Scalars and Vectors

Scalars and vectors are fundamental concepts in physics and mathematics that are essential for understanding motion and describing quantities in the physical world. They play a crucial role in the study of classical mechanics and are used to quantify various physical properties. Let's explore the definitions, characteristics, and differences between scalars and vectors.

Scalars: Scalars are physical quantities that have magnitude but no direction. In other words, scalars are described solely by a numerical value and a unit of measurement. Common examples of scalars include mass, temperature, time, speed, volume, and energy. When dealing with scalars, mathematical operations like addition, subtraction, multiplication, and division are straightforward, as they involve simple arithmetic. For example, if we consider the speed of a car, which is 60 kilometers per hour, the magnitude is 60, and the unit is kilometers per hour. Here, the direction of the car's motion is not considered; only the numerical value and unit are relevant (Figure 1.4).

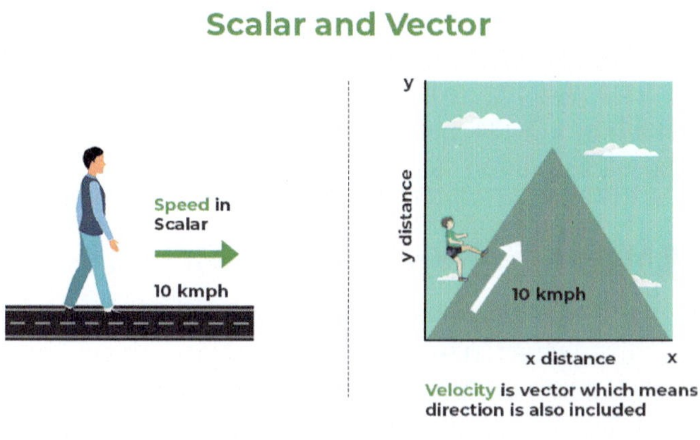

Figure 1.4. Difference between scalar and vector (Source: Anjali Shukla, Creative Commons License)

Vectors are a class of physical quantities characterized by their possession of both magnitude and direction. To provide a comprehensive description of a vector, it is necessary to explicitly indicate its magnitude, representing the extent or quantity, and its direction, indicating the orientation or spatial orientation. In graphical representation, vectors are depicted as arrows, where the arrow's length signifies the vector's magnitude.

In contrast, the arrow's direction signifies the vector's direction. Various examples of vectors encompass displacement, velocity, acceleration, force, and momentum. When engaging with vectors, mathematical operations encompass the processes of vector addition, vector subtraction, and scalar multiplication.

As an illustration, when examining the velocity of an airplane, which measures 400 kilometers per hour in an eastern direction, the magnitude of the velocity is 400, while the direction is east. To comprehensively describe the airplane's motion, it is essential to consider both the numerical value and the direction (Sekihara et al., 2004).

Tip

When analyzing motion, always remember to distinguish between scalars (quantities with magnitude only) and vectors (quantities with both magnitude and direction) to accurately describe the motion of objects.

This chapter aims to discuss the distinguishing characteristics of scalars and vectors. Several key distinguishing characteristics differentiate scalars from vectors:

a. Magnitude: Scalars have only magnitude, representing the physical quantity's size or quantity. Vectors have both magnitude and direction.
b. Representation: Scalars can be represented by a single numerical value and a unit of measurement. Vectors are represented graphically as arrows or symbolically using boldface letters or with a hat (^) above the letter (e.g., v or ⍰) (Ginchev et al., 2006).
c. Operations: Using regular algebraic rules, scalars can undergo ordinary arithmetic operations, such as addition and subtraction. Conversely, vectors require specific rules for vector addition and subtraction, considering both magnitude and direction.
d. Graphical Representation: Scalars are depicted on a number line or a graph with only one axis. Vectors are represented by magnitude and direction arrows, requiring a two-dimensional or three-dimensional coordinate system (Murphy, 1972).

Combining Scalars and Vectors: In many real-world scenarios, scalars and vectors are combined to describe complex physical quantities accurately. For instance, when calculating the total displacement of an object, both the scalar distance traveled, and the vector direction of motion need to be considered. This combination of scalars and vectors is fundamental to analyzing motion and forces in classical mechanics (Pachava et al., 2020).

1.5.2. Position, Displacement, and Distance

The comprehension of object motion in classical mechanics relies on the fundamental concepts of position, displacement, and distance. The spatial relationship and changes in the position of an object are elucidated in terms of its relative proximity to a reference point or a frame of reference. In this discourse, we shall examine and elucidate the definitions and distinctions of these significant notions (Suh, 2019).

Position is a term used to describe the spatial location of an object about a selected reference point, typically regarded as the origin of a coordinate system. In the context of one dimension, a solitary numerical quantity along a linear continuum denotes the concept of position. Conversely, numerical values corresponding to multiple axes represent the position in two or three dimensions. As an illustration, in the context of a car's motion along a linear path, its instantaneous location could be 100 meters relative to a designated initial reference point. In a two-dimensional context, the spatial coordinates of an aircraft can be denoted as (x, y) (Figure 1.5) (Mitiche et al., 1988).

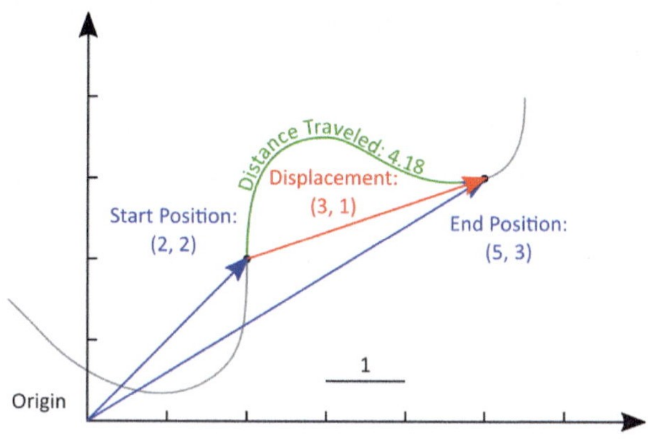

Figure 1.5. Difference between position, distance and displacement (Source: Stack Exchange, Creative Commons License)

Displacement is defined as the alteration in the spatial location of an entity, specifically denoting the transition from its original placement to its ultimate placement.

The quantity under consideration possesses vector characteristics due to its possession of both magnitude and direction. The magnitude of displacement refers to the minimum distance separating the initial and final positions, whereas the direction denotes the orientation from the initial position toward the final position (Chernetsov et al., 2008).

For example, if the vehicle above is displaced from its initial location to a point 150 meters distant, its displacement would be measured as 50 meters in the positive direction, specifically towards the right. In the context of two or three dimensions, the concept of displacement can be effectively conveyed through vectors. These vectors accurately depict the alteration in position by encompassing both the magnitude and direction of the change (Weiss et al., 2006).

Distance is a scalar quantity that denotes the overall length of the trajectory traversed by an object in the course of its motion. In contrast to displacement, distance is a measure that considers the specific route followed, irrespective of the direction. Distance is intrinsically non-negative and devoid of negativity, which signifies a scalar magnitude. As an illustration, in the scenario where the car traverses a curved road from its initial location to a final position 150 meters distant but follows a longer trajectory, the total distance covered could be approximately 180 meters. The measurement of distance solely takes into account the linear extent of the path traversed without considering the specific orientation or direction followed (Sugiyama & Yokoyama, 1996).

a. If an object moves in a straight line from one point to another, the displacement and distance are the same since the direction of motion and the path are identical.

b. If an object moves in a straight line but returns to its initial position, the displacement is zero, as there is no net change in position. However, the distance traveled is nonzero and corresponds to the total length of the path taken.

c. If an object moves along a curved path or changes direction during its motion, the displacement will be less than the distance traveled, as the displacement only considers the straight-line change in position (Gateno et al., 2004).

1.5.3. Speed and Velocity

Speed and velocity are fundamental principles within classical mechanics, which characterize the rate at which an object undergoes motion and the specific direction in which it travels. Although both terms are associated with the speed of movement, they possess separate definitions and interpretations. A comprehensive comprehension of the distinctions between average speed and velocity, along with instantaneous speed and velocity, is imperative for the precise analysis and depiction of object motion (Figure 1.6) (Yang et al., 2003).

The notion of average speed pertains to the quotient obtained by dividing the total distance covered by the total duration of travel. The notion of average speed pertains to a scalar magnitude that measures the total distance traversed by an entity during a specified time interval. The calculation divides the total distance covered by the time needed to traverse that distance. The mathematical representation denoting the concept of average speed is formally defined as:

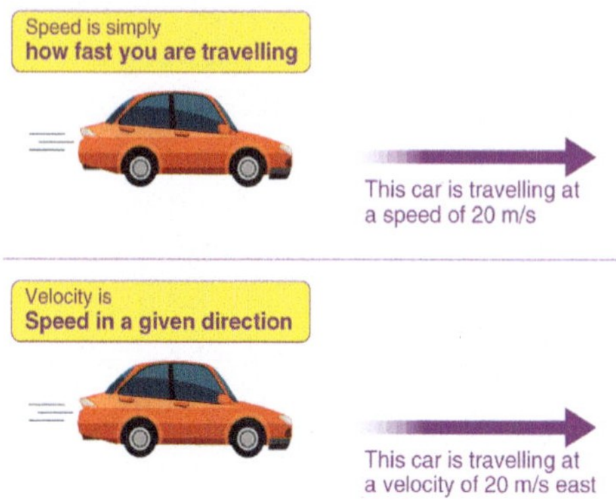

Figure 1.6. Difference between speed and velocity (Source: Byju's, Creative Commons License)

The average speed of an object can be determined by dividing the total distance traveled by the time taken.

The customary representation for average speed is distance per unit of time, commonly denoted as meters per second (m/s) or kilometers per hour (km/h). For instance, in the scenario where a vehicle covers a cumulative distance of 300 kilometers within a time frame of 5 hours, the resulting average speed can be calculated (Xu et al., 2020):

Average Speed = 300 km / 5 h or 60 km/h

The concept of average velocity pertains to a vector quantity that quantifies the displacement of an object within a designated time interval. Displacement refers to the alteration in spatial location between an initial point and a final point, encompassing both the magnitude and direction of the change. The mathematical expression representing average velocity is defined a:

Average Velocity = Displacement / Time Taken.

Since velocity is a vector quantity, it is imperative to incorporate both its direction and magnitude when representing it. For example, in the scenario where an object undergoes a displacement of 100 meters towards the east within a time interval of 2 seconds, the resultant average velocity of said object can be determined:

Average Velocity = 100 m east / 2 s = 50 m/s east

The direction is included to signify that the object moved eastward (Kafle et al., 2018).

The concept of instantaneous speed refers to the velocity of an object at a precise moment in time. The scalar quantity in question quantifies the instantaneous speed of an object, disregarding its direction. Calculating instantaneous speed involves determining the derivative of distance concerning time at a specific moment.

In practical applications, instantaneous speed is commonly determined using instruments such as vehicle speedometers or GPS-based tracking systems. These devices employ a continuous measurement technique to determine the distance traversed by an object within a brief time interval, subsequently utilizing this data to compute the object's instantaneous speed.

In this context, the "dt" denotes an infinitesimally small time interval.

As an illustration, let us examine an object undergoing motion along a linear path. The object's position can be described by the function x(t) = 5t^2 – 2t + 8, where "t" represents time measured in seconds, and "x" denotes the position measured in meters. The determination of the object's instantaneous velocity at any given time "t" can be achieved by differentiating the position function concerning "t" (SafarPour & Ghadiri, 2017):

Instantaneous Velocity = d(x(t)) / dt = d(5t^2 – 2t + 8) / dt = 10t – 2

1.5.4. Acceleration and Deceleration

1.5.4.1. Calculating Acceleration

Acceleration is a fundamental concept in classical mechanics that measures the rate of change of an object's velocity over time. It is a vector quantity, which means it includes both magnitude and direction. Acceleration occurs when an object's velocity changes, either by increasing (positive acceleration) or decreasing (negative acceleration, commonly known as deceleration).

The formula to calculate acceleration is:

Acceleration (a) = Change in Velocity (Δv) / Change in Time (Δt)

Acceleration is measured in units of distance per time squared, such as meters per second squared (m/s^2).

For example, if a car initially travels at a velocity of 20 m/s and then increases its velocity to 30 m/s over a time interval of 4 seconds, the acceleration can be calculated as follows:

Acceleration (a) = (30 m/s – 20 m/s) / 4 s = 2.5 m/s^2 The positive value of acceleration indicates that the car is speeding up.

1.5.4.2. Motion Graphs and Interpretation

Motion graphs are graphical representations of an object's motion, depicting its position, velocity, and acceleration as functions of time. There are three primary types of motion graphs: position-time graphs, velocity-time graphs, and acceleration-time graphs.

a. Position-Time Graphs: A position-time graph shows how an object's position changes over time. The object's position is plotted on the vertical axis, while time is plotted on the horizontal axis. The slope of a position-time graph represents the object's velocity, and the curvature of the graph indicates whether the object is accelerating or decelerating. A straight line on a position-time graph indicates constant velocity, while a curved line suggests a changing velocity. If the curve is steepening, the object is accelerating. Conversely, if the curve is flattening, the object is decelerating.

b. Velocity-Time Graphs: A velocity-time graph illustrates an object's velocity as a function of time. Velocity is plotted on the vertical axis, and time is plotted on the horizontal axis. The slope of a velocity-time graph represents the object's acceleration. A straight horizontal line on a velocity-time graph indicates constant velocity, as there is no change in velocity over time. An upward-sloping line represents positive acceleration (speeding up), and a downward-sloping line represents negative acceleration or deceleration (slowing down).

c. Acceleration-Time Graphs: An acceleration-time graph depicts an object's acceleration as a function of time. Acceleration is plotted on the vertical axis, and time is plotted on the horizontal axis. The slope of an acceleration-time graph indicates the rate of change of acceleration, which is known as jerk. A horizontal line on an acceleration-time graph indicates constant acceleration, while a sloped line represents changing acceleration. Positive slopes signify increasing acceleration, while negative slopes indicate decreasing acceleration or deceleration.

Interpreting motion graphs allows us to visualize an object's motion and gain insights into its behavior, whether it is moving at a constant speed, accelerating, or decelerating.

Acceleration is the rate of change of velocity over time and can be positive (speeding up) or negative (decelerating). Motion graphs, such as position-time, velocity-time, and acceleration-time graphs, provide valuable visual representations of an object's motion characteristics. By analyzing these graphs, we can better understand how an object's position, velocity, and acceleration change over time, aiding our comprehension of classical mechanics and kinematics (Radford et al., 2008).

1.5.5. Projectile Motion

1.5.5.1. Understanding Trajectories

The concept of projectile motion in classical mechanics is an intriguing phenomenon that characterizes the trajectory of an object launched into the atmosphere while being subjected to the force of gravity. When an object is projected at an angle relative to the horizontal plane, it traces a curved path called a trajectory. The trajectory of an object is influenced by the interplay between its initial velocity and the force of gravity exerted on it during its entire duration of motion. Treating the horizontal and vertical motions as separate entities is common practice in the context of projectile motion. Assuming negligible air resistance, the horizontal velocity of the object remains constant, while its vertical motion is subject to the force of gravity, resulting in a downward acceleration. The amalgamation of horizontal and vertical movements gives rise to a trajectory following a parabolic path (Figure 1.7) (Jimoyiannis & Komis, 2001).

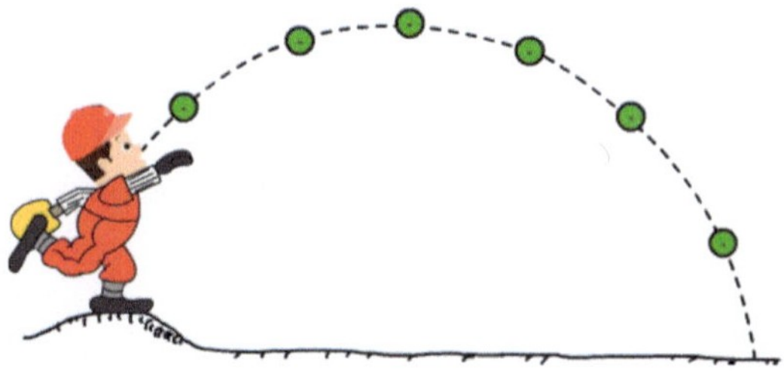

Figure 1.7. An example of projectile motion (Source: Bartleby, Creative Commons License)

1.5.5.2. Horizontal and Vertical Motions

A consistent horizontal velocity distinguishes the horizontal component of motion in projectile motion. The horizontal velocity of the object remains constant during its trajectory, assuming that no external horizontal forces, such as air resistance, are exerted on it. The principle above can be understood as a direct result of Newton's first law of motion, which asserts that an object in motion will persist in its motion with an unchanging velocity when no external forces are acting upon it. Consequently, the horizontal displacement achieved by a projectile exhibits a direct correlation with its horizontal velocity and the duration of its flight (Figure 1.8).

Did You Know?

Projectile motion is a classical mechanics concept used in various fields, including sports, engineering (e.g., ballistics), and astronomy (e.g., calculating orbits of celestial bodies)

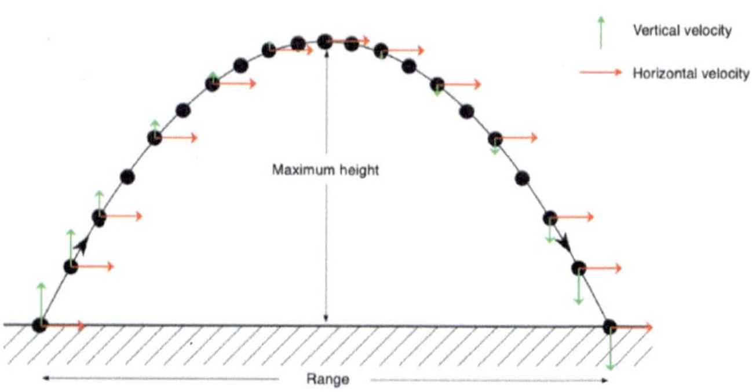

Figure 1.8. Horizontal and vertical components of projectile motion (Source: Science Ready, Creative Commons License)

The formula can be utilized to calculate the horizontal distance traveled, also known as the range (Kawase et al., 2011):

Range = Horizontal Velocity × Time of Flight

The vertical component of projectile motion is governed by the force of gravity, resulting in a consistent downward acceleration of the object. The acceleration resulting from the gravitational force, commonly represented as "g," is approximately 9.81 m/s^2 near the Earth's surface. The vertical velocity of an object changes as it ascends or descends. At the maximum point of the trajectory, commonly referred to as the apex, the vertical velocity reaches zero before initiating descent due to the force of gravity. The duration required to reach the highest point can be determined by employing a calculation method (Patra & Haldar, 2022):

Time to Apex = Vertical Velocity / Acceleration due to Gravity (g)

Similarly, the total time of flight (the time taken for the object to return to the same horizontal level from which it was launched) can be calculated using:

Total Time of Flight = 2 × Time to Apex

The maximum height (vertical displacement) reached by the projectile can be determined using (Sansón et al., 2012):

Maximum Height = (Vertical Velocity at Apex)^2 / (2 × Acceleration due to Gravity) CProjectile motion is prevalent in various real-world scenarios, such as the motion of a football during a kick, the trajectory of a launched rocket, or the flight of a baseball after being hit by a bat. Understanding the separation of horizontal and vertical motions in projectile motion allows us to analyze and predict the path of a projectile accurately, making it a fundamental topic in the study of classical mechanics (Konno & Ohmachi, 1998).

1.5.6. Uniform Circular Motion

1.5.6.1. Centripetal and Centrifugal Forces

Uniform circular motion is a term used to describe the movement of an object as it travels in a circular path at a consistent velocity. The object's velocity remains constant, but its trajectory undergoes continuous alterations, resulting in a circular trajectory. In the present context, two discernible forces, commonly referred to as centripetal and centrifugal forces, are observed to be operative (Figure 1.9) (Montoya, 1999).

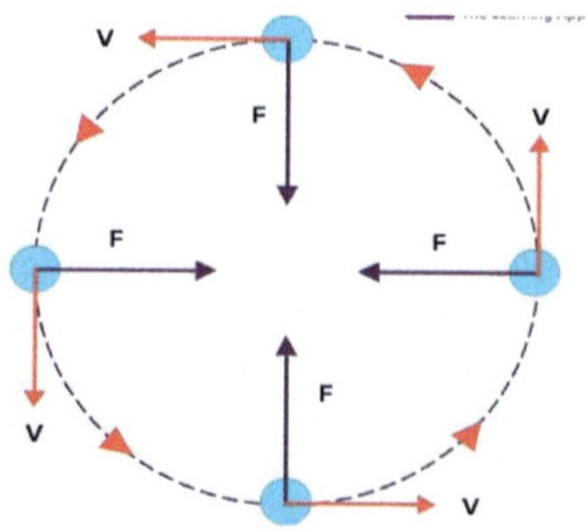

Figure 1.9. Uniform circular motion (Source: Byju's, Creative Commons License)

Centripetal force refers to the collective force acting towards the center of a circle, which is responsible for maintaining an object's uniform circular motion. It is important to note that centripetal force is not independent but a descriptive term denoting the net force in this context. The centripetal force is necessary to consistently alter an object's direction, thereby maintaining its trajectory along a circular path. Centripetal force is often generated by tension in a string, gravitational attraction, friction between tires and the road, or any other force that is oriented toward the center of the circular motion. An illustrative instance involves a car executing a turn, wherein the interaction of friction between the car's tires and the road surface furnishes the requisite centripetal force to sustain the car's motion along a curved trajectory (Figure 1.10) (Cook et al., 2007).

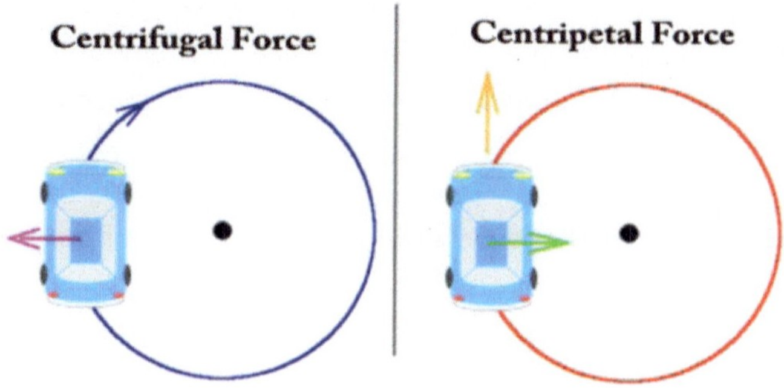

Figure 1.10. Difference between centripetal and centrifugal force (Source: Assignment Point, Creative Commons License)

Centrifugal force, in contrast to centripetal force, is commonly characterized as a "fictitious force" due to its apparent outward action from the center of the circle, resulting in the repulsion of objects. Nevertheless, it is imperative to comprehend that centrifugal force is not a genuine force but rather a perceived force encountered by individuals within a rotating frame of reference. The phenomenon occurs due to the object's inertia, which causes it to resist deviating from its linear trajectory while traversing a curved path. To summarize, the centripetal force is directed inward and maintains the motion of an object in a uniform circular path. On the other hand, centrifugal force is a perceived force encountered by individuals within a rotating frame of reference. Still, it does not correspond to an actual force exerted on the object (Barwell, 2014).

1.5.6.2. Angular Velocity and Acceleration

Angular velocity measures how quickly an object's angular displacement changes over time. The measurement pertains to the object's angular velocity, denoting the rate at which it rotates or traverses an angle. Angular velocity is a vector quantity represented by the symbol "ω" (omega).

The formula utilized for the determination of angular velocity is:

Angular Velocity (ω) = Change in Angular Displacement / Change in Time

Angular velocity is often given in radians per second (rad/s), and its value can be either positive or negative, depending on the direction in which the rotation occurs (clockwise or counterclockwise). Acceleration of the Rotational Axis: Conversely, angular acceleration refers to the rate at which angular velocity shifts about time and quantifies this shift. It accounts for the rapidity with which the object's angular velocity is altering. Another type of vector quantity, angular acceleration, is represented by the symbol " (alpha), which stands for "alpha" (Yaghobi, 2016).

The following equation can be used to compute angular acceleration:

Change in angular velocity divided by change in time is the formula for angular acceleration.

The radians per second squared (rad/s2) is the unit of measurement for angular acceleration, which indicates whether the item is experiencing an increase or a decrease in its rotational speed.

The term "uniform circular motion" describes items moving in a circular direction while keeping a constant velocity throughout the journey. The resultant force that acts towards the center of an item is known as the centripetal force, which is responsible for maintaining the object's circular trajectory. On the other hand, centrifugal force is a perceived force encountered within a rotating frame of reference. This force acts in the opposite direction of the rotation. Angular velocity is a metric that measures the speed at which an object rotates.

In contrast, angular acceleration is a term that describes the rate at which an item's angular velocity changes over time. A thorough understanding of these concepts is necessary to analyze and use circular motion in various fields of study, such as mechanics, engineering, and celestial dynamics (Rodriguez-Maldonado, 2019).

1.6. FORCES AND MOTION

Dynamics is a subfield within classical mechanics investigating the interplay between forces and the resulting motion. This study's primary objective is to comprehend how forces exerted on objects impact their acceleration and subsequent motion. Applying Newton's three laws of motion allows for establishing a comprehensive framework within the field of dynamics, which facilitates the description of object behavior when subjected to external forces. The underlying principle enables the anticipation and examination of the movement of various entities, ranging from mundane items to celestial entities. Consequently, dynamics assumes a pivotal role in contemporary physics and engineering (Moo et al., 2018).

1.6.1. Force Types and Fundamental Forces in Nature

Forces are fundamental interactions that govern the behavior of matter in the universe. In nature, there are four fundamental forces, each with distinct characteristics and ranges of influence. Let's explore the four fundamental forces (Figure 1.11) (Field et al., 1985):

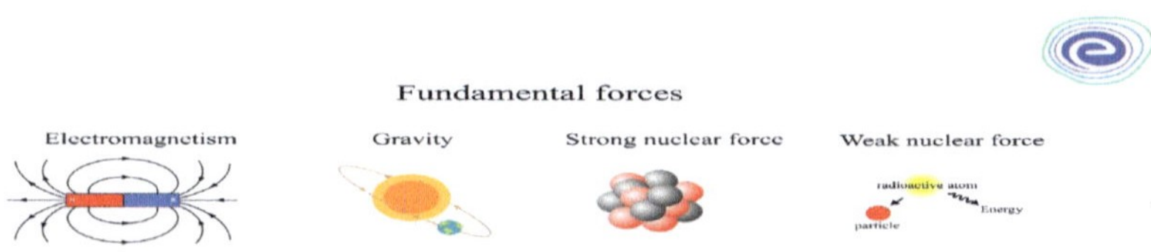

Figure 1.11. Fundamental forces of nature (Source: Anum, Creative Commons License)

1. Gravitational Force: There is consensus among scientists and researchers that gravitational force is one of the universe's most fundamental forces. The gravitational attraction phenomenon, which is essentially driven by the force of gravity, plays an essential role in controlling the motion of celestial entities such as planets, stars, and galaxies. Gravity is the primary force that governs the gravitational attraction phenomenon. When two things with mass are brought together, a gravitational force is created that is directly proportional to the product of those masses and inversely proportional to the square of the distance that separates their centers. This force is called gravitation. The phenomena of gravitational force can be explained by Sir Isaac Newton's law of universal gravitation, which postulates that every particle in the universe exerts an attractive force on every other particle. This law explains the gravitational force phenomenon. The amplitude of this force has been shown to demonstrate a direct proportionality to the product of the masses of the objects involved while at the same time displaying an inverse proportionality to the square of the distance that separates them from one another. Gravity, while the weakest of the fundamental forces, has a wide range and is responsible for the dynamics of astronomical scales, such as celestial bodies' motion and galaxies' formation. This is despite gravity being the least powerful of the fundamental forces (Mintzberg, 1991).
2. Electromagnetic Force: The electromagnetic force is a fundamental force that regulates the interactions among charged particles, specifically electrons and protons. It is accountable for diverse phenomena, encompassing the conduct of atoms, molecules, and electromagnetic radiation. The electromagnetic force can be divided into two distinct components: the attractive force that occurs between charges of opposite polarity and the repulsive force that arises between charges of the same polarity. Like the force of gravity, the electromagnetic force also adheres to the principle of an inverse square law, wherein the magnitude of the force diminishes proportionally to the square of the distance separating the charges. As postulated by James Clerk Maxwell, the laws of electromagnetism provide a comprehensive framework for understanding the phenomenon of electromagnetism, effectively merging the principles of electricity and magnetism

into a cohesive and unified theory. The electromagnetic force is of utmost significance in chemistry, electronics, and the operational mechanisms of commonplace artifacts, encompassing magnets and electrical apparatus (Ziefle, 2016).

3. Strong and Weak Nuclear Forces: The interactions occurring at the subatomic level within the atomic nucleus are attributed to the strong and weak nuclear forces. The strong nuclear force, referred to as the strong interaction or strong force, is the fundamental force responsible for the cohesive binding of protons and neutrons within the atomic nucleus. The phenomenon in question involves an attractive force that surpasses the electrostatic repulsion arising from the positive charges of protons. The strong nuclear force exhibits significant strength when acting over extremely short distances, yet its potency diminishes rapidly as the distance between interacting particles increases. Consequently, the force's efficacy is confined solely to the confines of the atomic nucleus.
4. Conversely, the weak nuclear force, also known as the weak interaction, is accountable for specific forms of radioactive decay and interactions involving neutrinos. The weak force exhibits significantly lower strength than the electromagnetic and strong forces, and a notably limited range characterizes it. The strong and weak nuclear forces collectively exert significant influence on the stability and dynamics of atomic nuclei, making them indispensable for comprehending the phenomena taking place in stars and nuclear reactions (Antoniadis et al., 2011).

1.6.2. Free Body Diagrams and Force Analysis

Free body diagrams (FBDs) and force analysis are fundamental techniques employed in classical mechanics to visually depict and examine the forces exerted on an entity. A free-body diagram is a rudimentary, proportionally reduced illustration of an object wherein the object is separated from its immediate environment, and the vectors representing all the forces acting upon it are depicted.

To construct a free-body diagram, it is customary to represent the object under consideration as a singular point or a basic geometric form, such as a rectangle or a circle. The object is subject to various forces, which are graphically depicted as arrows emanating from a central point and aligned in the direction of each respective force. Each arrow's length represents the force's magnitude, while the arrow's orientation indicates the force's direction (Kurniawan et al., 2023).

After constructing the free body diagram, the force analysis process entails identifying and examining each force depicted in the diagram. This entails the identification of the specific category of force (e.g., gravitational, frictional, normal, tension, etc.), quantifying its intensity, and determining its orientation.

The analysis of forces aids in comprehending the impact of forces on an object's motion and determining whether the object is in a state of equilibrium, where the net force is zero, or experiencing acceleration, where the net force is nonzero. Through free-body diagrams and force analysis, individuals in engineering, physics, and academia can effectively examine and forecast the responses of objects exposed to diverse forces. This practice enables a more profound comprehension of mechanics while facilitating the resolution of problems across a broad spectrum of practical situations (Atilola et al., 2014).

1.6.3. Applications of Newton's Laws

Newton's laws of motion find numerous applications in our daily lives and various fields of science and engineering. Let's explore three essential applications:

1. Frictional Forces: When one surface tries to slide or move relative to the other, this might cause forces of friction to develop between the two surfaces. Applying Newton's first law of motion is important to remember when attempting to comprehend the idea of friction. It depends on the coefficient of friction between the interacting surfaces as to the magnitude of the force required to commence the motion of an object by overcoming the resistance that friction presents. Newton's second law describes the force required to combat friction and sustain the continuous motion of an item at a constant velocity. This law was named after Newton himself. It is possible to give a mathematical representation of the frictional force in the form of an equation (Thornton & Sokoloff, 1998):

Force of Friction = Coefficient of Friction × Normal Force.

2. Tension in Ropes and Cables: Understanding tension in ropes and cables is fundamental in engineering and construction. When

an external force is exerted on an object suspended by a rope or cable, the tension in the rope propagates the force uniformly along its entire length, thereby enabling the object to maintain a state of equilibrium.

As per Newton's third law of motion, the tension exerted by the rope or cable in an upward direction must be equivalent in magnitude but opposite to the object's weight, exerting a downward force. This condition guarantees that the object remains in a state of rest or moves with a consistent velocity. Comprehending tension is of utmost importance in practical applications, particularly in designing and constructing suspension bridges, cranes, elevators, and other structures that rely heavily on cables and ropes to support substantial loads (Patil et al., 2022).

3. Circular Motion and Centripetal Forces: The laws formulated by Sir Isaac Newton are also utilized in comprehending circular motion and the concept of centripetal forces. When an object undergoes circular motion, it encounters a centripetal force oriented toward the circle's center. Newton's second law states that the force is accountable for the object's acceleration toward the circle's center. The expression that represents the centripetal force necessary to maintain the motion of an object in a circular path is defined a (Firdawati et al., 2021):

Centripetal Force = (Mass × (Linear Speed)^2) / Radius of the Circular Path

Centripetal forces are of considerable importance in diverse practical situations. This phenomenon can be observed when a vehicle executes a turn. In this scenario, the interaction between the tires and the road generates the necessary centripetal force to maintain the car's trajectory along the curved path. The phenomenon of circular or elliptical orbits in celestial mechanics is attributed to the gravitational force exerted by planets and their stars (Sakalayen et al., 2021).

The laws of motion formulated by Sir Isaac Newton possess significant utility in comprehending the dynamics of forces across diverse scenarios. Frictional forces exert an impact on various routine activities encountered in daily life. The presence of tension in ropes and cables holds significant importance within the domains of engineering and construction. Furthermore, the principles of circular motion and centripetal forces provide a comprehensive understanding of the dynamics governing the movement of objects along curved trajectories. The applications showcased in these examples illustrate Newton's laws' extensive and varied importance across various disciplines, establishing them as fundamental principles within classical mechanics and playing a significant role in scientific and technological advancements (Del Giudice et al., 2016).

1.6.4. Applying Newton's Laws in Real-World Scenarios

Newton's laws of motion are ubiquitous in real-world scenarios, offering insights into the behavior of various objects and systems. Here are three significant applications of Newton's laws in everyday life:

1. Motion of Vehicles: Newton's laws play a crucial role in understanding the motion of vehicles, whether on land, in the air, or on water. When a vehicle accelerates, decelerates, or turns, it experiences various forces, and Newton's laws provide a framework for analyzing these dynamics.

   For instance, when a car accelerates forward, Newton's second law explains that the net force acting on the car is the product of its mass and acceleration. The engine generates a force that propels the car forward, overcoming frictional forces and air resistance. During a turn, the car experiences centripetal forces that keep it moving in a circular path. These forces result from the interaction between the tires and the road. The car's tires provide the necessary friction to generate the centripetal force required to navigate the curve safely (Husna & Fajar, 2022).

2. Falling Bodies and Projectiles: The motion of falling bodies and projectiles is governed by Newton's laws of motion, especially the first and second laws. When an object is dropped or thrown vertically, it experiences the force of gravity, causing it to accelerate downward at a constant rate (approximately 9.81 m/s^2 near the Earth's surface).

   Newton's second law describes this behavior, relating the object's mass to its acceleration. Objects thrown at an angle for projectile motion experience horizontal and vertical motions. The horizontal component of motion remains constant due to Newton's first law, while the vertical component experiences constant acceleration due to gravity. As a result, the path followed by a projectile is a parabolic trajectory (Sun et al., 2018).

3. Forces in Simple Machines: Simple machines, such as levers, pulleys, inclined planes, and screws, are fundamental tools that facilitate mechanical work. Newton's laws underpin the principles behind the operation of these machines. For instance, in a lever system, the effort applied to one end is transmitted to the load on the other, and the fulcrum provides the pivot point. The balance of forces at the fulcrum follows Newton's first law, ensuring equilibrium. Additionally, the lever's mechanical advantage, which amplifies force or distance, relies on Newton's second law. Similarly, pulleys use Newton's laws to redistribute forces and make lifting heavy objects easier. The arrangement of ropes and pulleys distributes the load force over multiple ropes, reducing the required effort force (Halloun, 1998).

Newton's laws of motion are invaluable tools for understanding the behavior of various objects and systems in everyday life. From the motion of vehicles and falling bodies to the operation of simple machines, these laws provide essential insights into how forces shape the world around us. Engineers, physicists, and designers use these laws to create safer and more efficient technologies, making Newton's laws a cornerstone of classical mechanics and essential to the progress of science and engineering (Shekhar et al., 2014).

1.7. ENERGY AND WORK

Energy and work are fundamental concepts in physics that exhibit a close interrelationship. Energy can be defined as the inherent ability of a given system to perform work, which in turn refers to transferring energy from one system to another by applying force over a measurable distance. The completion of work occurs when an external force is applied to an object, resulting in its displacement over a specific distance in the same direction as the force.

The equation for calculating the work done on an object is given by multiplying the applied force by the distance traveled in the direction of the force. Energy is present in diverse manifestations, including kinetic energy, which is linked to the motion of objects, potential energy, which is associated with the position of objects; and thermal energy, which is connected to the temperature of objects.

The principle of energy conservation posits that the total amount of energy within a closed system remains constant over time, with no creation or destruction of energy occurring. Instead, energy transforms, transitioning from one form to another. A comprehensive grasp of the principles of energy and work is imperative for examining diverse physical phenomena, encompassing mechanical systems, heat transfer, and the dynamics of particles in various fields of study. The concepts of energy and work are fundamental in physics and are closely interconnected. The concept of energy can be described as the intrinsic capacity of a specific system to carry out mechanical or physical tasks (Figure 1.12) (Owens et al., 2016).

On the other hand, the term "work" denotes the act of transferring energy from one system to another by exerting force over a defined distance. Within physics, the concept of work is characterized as the outcome of multiplying a force exerted on an entity by the displacement experienced by the said entity in the same direction as the force.

Figure 1.12. Work is given by the equation Fdcosθ, where d is the horizontal displacement over which the force acts and θ is the angle between the force and distance (Source: Byju's, Creative Commons License)

The equation used to determine the amount of work performed on an object involves multiplying the force exerted on the object by the distance over which the object is displaced in the same direction as the applied force.

Energy exists in various forms, encompassing kinetic energy, which is correlated with the movement of objects, potential energy, which is related to the spatial arrangement of objects; and thermal energy, which is tied to the temperature of objects.

The principle of energy conservation states that the total energy within a closed system remains constant throughout time. Based on this principle, the conservation of energy states that energy is neither created nor destroyed but rather undergoes conversion from one form to another. A thorough understanding of the fundamental concepts of energy and work is essential for analyzing a wide range of physical phenomena, including mechanical systems, heat transfer, and the dynamics of particles in different domains of physics (Methfessel et al., 1992).

1.7.1. Definition of Energy

Energy is a fundamental and essential physical quantity that measures a system's capacity to carry out work or induce a modification. The property in question is classified as a scalar, indicating that it possesses magnitude but lacks direction. Energy is present in diverse manifestations, including kinetic energy, which is linked to the movement of an object, potential energy, which is associated with the position or arrangement of a system, thermal energy, which is connected to temperature and the transfer of heat; and numerous other forms. The fundamental principle of energy conservation posits that energy is a conserved quantity, meaning it cannot be generated or annihilated. Instead, energy can solely undergo processes of transfer or conversion, wherein it transitions from one manifestation to another (Izquierdo et al., 2010).

1.7.2. Kinetic Energy and the Work-Energy Theorem

Kinetic energy is a type of energy linked to an object's movement. The discussed concept is of utmost importance in classical mechanics as it pertains to quantifying an object's energy resulting from its velocity.

Kinetic Energy (KE) = ½(m * v^2) Therefore, an object with a higher velocity or greater mass will exhibit greater kinetic energy. The Work-Energy Theorem establishes a correlation between the amount of work exerted on an object and the resulting alteration in its kinetic energy. As per the theorem, the net work (W) exerted on an object is equivalent to the alteration in its kinetic energy. In mathematical terms, it is expressed as (Pili

& Violanda, 2021):

Net Work (W) = Change in Kinetic Energy (ΔKE)

The work-energy theorem comprehensively explains how external forces acting on an object can cause changes in its kinetic energy. When a net external force acts upon an object, it experiences acceleration, and work is performed on the object to modify its velocity. This study demonstrates a modification in the quantity of kinetic energy. If the net work is positive, it increases kinetic energy, causing the object to experience accelerated motion. On the contrary, in cases where the net work is negative, there is a decrease in the overall kinetic energy, leading to deceleration or a decrease in velocity (Bonanno et al., 2016).

Utilizing the work-energy theorem is a highly efficient method for examining various physical systems, including simple collisions and complex mechanical processes. Examining forces, energy, and motion facilitates our understanding of the relationships between these principles, providing a holistic perspective on how objects respond to external stimuli and how energy is transferred and transformed in different situations (Lawson & McDermott, 1987).

1.7.3. Potential Energy

An object possesses a sort of energy known as potential energy due to its spatial placement or configuration of the environment in which it is situated. The energy held within a thing is referred to as its potential energy, a term used to define it.

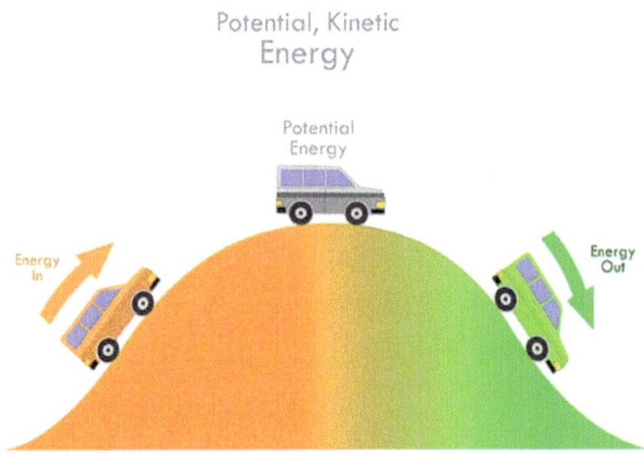

Figure 1.13. Potential and kinetic energy (Source: Tara Energy, Creative Commons License)

This energy can be turned into a variety of other forms of energy, one of which is kinetic energy. This occurrence takes place whenever the thing in question moves or is subjected to the influence of an outside force or forces. The following talk will concentrate on two of the many different kinds of potential energy (Figure 1.13) (Lorenz, 1955).

Within physics, gravitational potential energy is essential to the whole framework. The energy that an object holds as a direct result of its location inside a gravitational field is referred to as its "gravitational potential energy," and it is denoted by the phrase "gravitational potential energy" (GPE). The idea of gravitational potential energy, abbreviated as GPE, refers to the energy already present in an object as a direct result of how it is spatially arranged about a gravitational field. The behavior witnessed can be explained by the gravitational force of attraction between the object in question and the Earth or possibly any other celestial body. The following equation can be used to determine the gravitational potential energy (GPE) of an item, which is denoted by its mass, "m," and which is positioned at a vertical displacement, "h," from a chosen reference point, which is often the surface of the Earth:

Gravitational Potential Energy = m * g * h

When an object is raised in opposition to the force of gravity, work is performed on the object, increasing its gravitational potential energy. When an object descends or undergoes downward motion, its gravitational potential energy (GPE) is converted into kinetic energy. This cyclic process persists as potential energy interconverts with kinetic energy. The utilization of gravitational potential energy holds considerable importance in diverse practical contexts, exemplified by its application in hydroelectric power plants. In this context, water is stored at an elevated position and subsequently released, converting its gravitational potential energy into kinetic energy, which is harnessed to produce electricity. The concept of elastic potential energy refers to the energy stored in an elastic material when deformed. Elastic potential energy refers to the energy stored within an object due to its deformation or compression. It is commonly linked with materials with elastic properties, such as springs and rubber bands. When an elastic material undergoes deformation through stretching or compression, mechanical work is exerted on it, accumulating potential energy within the material (Schlegel, 2003).

The formula for calculating the elastic potential energy (EPE) in a spring, which has a spring constant denoted as "k" and is either stretched or compressed by a displacement denoted as "x" from its equilibrium position, is as follows:

Elastic Potential Energy (EPE) = 0.5 * k * x^2

The magnitude of the elastic potential energy stored in the spring is directly proportional to the spring's displacement from its equilibrium position. Upon the release of the spring, the potential energy that was previously stored within it undergoes a conversion into kinetic energy, resulting in the spring reverting to its initial configuration. Elastic potential energy finds application in various devices and contexts, including mechanical systems, shock absorbers, and recreational items such as slingshots and catapults. The concept of potential energy holds significant importance in the field of physics as it denotes the energy that is stored within an object as a result of its specific position or configuration. The concept of gravitational potential energy is contingent upon the spatial arrangement of an object within a gravitational field. In contrast, elastic potential energy is intrinsically linked to the alteration in shape or form of elastic substances. The comprehension of these different manifestations of potential energy enables us to examine and forecast the actions of diverse systems and entities, thereby facilitating the development and enhancement of numerous pragmatic applications in engineering, technology, and everyday existence (Topping, 1927).

1.7.4. Conservation of Mechanical Energy

One of the most fundamental ideas in physics is the principle of conservation of mechanical energy. Given that an isolated system is only susceptible to conservative forces like gravity and elastic forces, the statement proposes that the total amount of mechanical energy does not change. This is because the system is not exposed to any other forces. The contribution of kinetic energy (KE) and potential energy (PE) to the overall mechanical energy of a closed system must be considered before arriving at an accurate estimate of the total mechanical energy (Fahrunnisa et al., 2021).

When a piece of matter is subjected to a conservative force field, such as the force of gravity, the object goes through a transition in which its mechanical energy changes from kinetic to potential states. As the object moves through the force field, the transformation described above happens to it. In the scenario of an object going through the motions of free fall, it can be shown that the acceleration that the item undergoes due to the force of gravity contributes to an increase in the object's kinetic energy. As an object descends downward to the point of lower vertical elevation, there is a corresponding decrease in its potential energy. This takes place simultaneously. However, suppose no non-conservative forces are acting on the system from the outside, such as air resistance or friction. In that case, the total energy, which accounts for kinetic and potential energy, is unaffected and stays the same throughout the motion (Pierratos & Polatoglou, 2017).

The oscillating oscillation of a pendulum is one of the best examples that can be used to illustrate the concept of the principle of conservation of mechanical energy. The pendulum demonstrates an oscillatory motion in which its mechanical energy alternates between kinetic energy, which manifests as maximum velocity at the lowest points of its trajectory, and potential energy, which manifests as maximum elevation at the highest points. This results in the pendulum swinging back and forth between these two states. The fact that the sum of the mechanical energy present in a system does not change regardless of its orientation exemplifies the concept of conservation. Establishing correlations between energies at different locations and forecasting the behavior of objects under the impact of conservative forces are two of the primary benefits that can be derived from applying the theory of mechanical energy conservation to the study of specific physical systems. Because it enables a comprehensive understanding of the underlying dynamics that control mechanical systems, the application of this instrument substantially impacts various subjects, including engineering, physics, and astronomy. This is because it enables the fields to comprehend fundamental dynamics better. Despite this, it is essential to consider the influence of non-conservative forces when investigating real scenarios in which the concept of the conservation of mechanical energy may not be applicable (Halilovic et al., 2022).

1.7.5. Power and its Measurement

The rate at which work is completed, or the rate at which energy is transmitted or transformed, is referred to as a system's power. It provides a numerical representation of the rate at which energy is either consumed or produced. Understanding the concept of power is essential to understanding the effectiveness and functionality of various processes and technologies.

The quantity of work (W) that is completed or the amount of energy (E) that is transported or transformed in a certain length of time (t) is the mathematical definition of power (P). The following mathematical expressions can be used to express it (Sidhu, 1999):

Power (P) = Work Done (W) / Time Taken (t)

Power (P) = Energy Transferred or Converted (E) / Time Taken (t)

In the International System of Units (SI), the watt (W) is the accepted standard power unit. It is the amount of work or energy transferred per second, equivalent to one joule. The unit of power known as the watt is typically employed to quantify relatively modest amounts of power. However, in practical applications, larger quantities of power are often measured using kilowatts (kW) or megawatts (MW).

The measurement of power holds significant importance across diverse disciplines, ranging from engineering and physics to practical everyday contexts. Assessing the operational effectiveness of electrical devices, motors, and generators in electrical systems heavily relies on power measurement. Power measurements are crucial in evaluating the efficiency of engines and machines within mechanical systems (Blackman & Tukey, 1958). Accurate measurement of power necessitates the utilization of specialized instruments. Wattmeters and power analyzers are commonly employed in electrical power measurement. Dynamometers and other devices are utilized in mechanical systems to quantify power output. The progression of technology has resulted in the development of advanced and precise power measurement instruments, which have empowered engineers and scientists to optimize energy consumption, enhance efficiency, and elevate the performance of diverse systems and devices (Meier & Schroder, 1984).

1.7.6. Work-Energy Applications in Real-Life Systems

Energy Conversion and Efficiency: Work-energy principles find extensive applications in energy conversion systems, where different forms of energy are transformed from one to another. For instance, in a power plant, mechanical work is done to spin turbines, converting kinetic energy into electrical energy. The work-energy theorem helps analyze the efficiency of this conversion process.

Efficiency is a vital metric in energy conversion systems, representing the ratio of useful output to total input energy. It can be expressed as:

Efficiency = (Useful Output Energy / Total Input Energy) * 100

Engineers can employ the work-energy theorem to compute the energies associated with the input and output of a process, evaluate losses attributable to phenomena such as friction and heat, and enhance the system's performance to attain greater efficiency. Comprehending the fundamental principles of energy conversion and efficiency is imperative in pursuing sustainable and environmentally conscious energy solutions, waste reduction, and optimal utilization of accessible resources (Andal & Jayapal, 2022).

1.8. MOMENTUM AND COLLISIONS

Momentum is a fundamental principle in physics that quantifies the amount of motion inherent in an object. "momentum" refers to the mathematical product of an object's mass and velocity. The mathematical representation of momentum (p) is given by the equation p = m * v, where "m" represents the mass of the object and "v" represents its velocity. Momentum is classified as a vector quantity, signifying that it possesses both magnitude and direction (Adams et al., 2003). The concept of momentum is of significant importance in the context of collisions. In the event of a collision between objects, the principle of conservation of momentum dictates that the sum of the momenta of all objects involved remains constant, provided that no external forces are exerted on the system. The principle above is commonly called the law of conservation of momentum. Collisions can be categorized into two distinct types: elastic collisions, characterized by the preservation of kinetic energy, and inelastic collisions, characterized by the dissipation of a portion of the kinetic energy. The comprehension of object behavior during interactions is significantly enhanced by studying momentum and its conservation in collisions. This concept is important in various disciplines, including mechanics, engineering, and ballistics (Ültay & Alev, 2017).

1.8.1. Definition of Momentum

The concept of momentum is a fundamental principle in physics, quantifying the amount of motion inherent in an object. The quantity under consideration possesses both magnitude and direction, qualifying as a vector. The concept of momentum can be defined as the mathematical product of an object's mass and velocity. The mathematical expression for momentum (p) is given by the product of mass (m) and velocity (v), as shown below: $p = m \times v$

The momentum of an object is contingent upon both its mass and velocity. Thus, an object with a greater mass or higher velocity will exhibit a greater momentum (Chaves, 2012). The direction of an object's momentum is congruent to the direction of its velocity. In the context of linear motion, the momentum of an object is deemed positive when it exhibits movement in the positive direction. At the same time, it is regarded as negative when it demonstrates movement in the negative direction. The unit of momentum in the International System of Units (SI) is denoted as kilogram-meter per second (kg·m/s). The provided scenario relates to the measurement of momentum, defined as the multiplication of an object's mass by its velocity. In the present scenario, the entity under consideration exhibits a mass of one kilogram and is observed to be in motion with a velocity of one meter per second (Szabados, 2009).

1.8.2. Impulse and the Impulse-Momentum Theorem

Impulse and the impulse-momentum theorem are important concepts in physics that relate to changes in an object's momentum resulting from external forces acting on it. Impulse is the product of the force applied to an object and the time interval over which the force acts. Mathematically, impulse (J) can be expressed as (Denny et al., 2020):

Impulse (J) = Force (F) × Time Interval (Δt)

The impulse-momentum theorem states that an object's momentum change equals the impulse applied to it. In equation form, the impulse-momentum theorem is written as:

Change in Momentum (Δp) = Impulse (J)

This theorem highlights the connection between force, time, and momentum. If an object experiences a constant force over a specific time interval, the impulse applied to it will change momentum. A crucial feature of the impulse-momentum theorem is that it allows for analyzing interactions that occur over a short period, such as collisions. When the time interval approaches zero (infinitesimal time), the impulse-momentum theorem defines instantaneous force in calculus (Figure 1.14) (Patel et al., 2018).

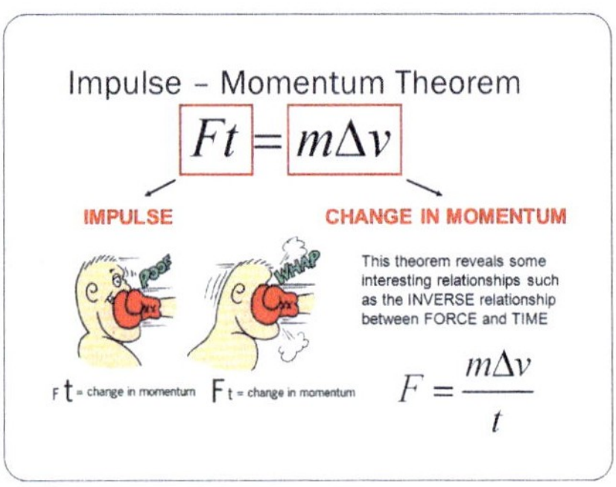

Figure 1.14. Momentum and impulse (Source: Quizlet, Creative Commons License)

Impulse is also related to the concept of average force. For an object subjected to a varying force over a given time interval, the average force (F_avg) can be calculated as:

Average Force (F_avg) = Impulse (J) / Time Interval (Δt)

The concept of impulse holds significant importance in sports and safety. In various sports, such as baseball, soccer, and basketball, athletes endeavor to exert an impulse on the ball to alter its momentum, resulting in forceful shots or precise passes. Applying the impulse-momentum theorem in safety features such as airbags and automobile crumple zones prolongs the collision duration. It mitigates the magnitude of the force exerted on occupants during accidents, diminishing the potential for harm (Pride et al., 1998).

1.8.3. Conservation of Momentum

The principle of momentum conservation is fundamental in physics that applies to analyzing collisions between objects. Based on this fundamental principle, the aggregate momentum of a closed system remains conserved both before and after a collision event, provided that no external forces are exerted upon the system. In essence, the principle of conservation of total momentum holds throughout the interaction, irrespective of the specific nature of the collision. This discussion examines the two primary categories of collisions: elastic and inelastic (Gartland & Slagsvold, 1975).

Elastic Collisions: In elastic collisions, momentum, and kinetic energy conservation principles are upheld. When two objects undergo an elastic collision, they exhibit a rebounding behavior without experiencing any dissipation of kinetic energy. This implies that the sum of the initial kinetic energies is equivalent to that of the final kinetic energies in the collision context. In the context of mathematics, the principle of conservation of momentum can be mathematically expressed for an elastic collision involving two objects with masses denoted as "m1" and "m2," initial velocities as "v1i" and "v2i," and final velocities as "v1f" and "v2f." (Bray & Eldridge, 2018):

m1 * v1i + m2 * v2i = m1 * v1f + m2 * v2f

The conservation of kinetic energy in an elastic collision can be represented as:

0.5 * m1 * (v1i)^2 + 0.5 * m2 * (v2i)^2 = 0.5 * m1 * (v1f)^2 + 0.5 * m2 * (v2f)^2

Elastic collisions are frequently observed in both macroscopic and microscopic systems. Elastic collisions encompass various scenarios, such as the collision between billiard balls, the collision between molecules within a gas, and certain subatomic interactions. In the scenarios above, the objects in question demonstrate a lack of enduring distortion and do not undergo any dissipation of kinetic energy throughout the collision event (Cape et al., 1989).

The phenomenon of inelastic collisions is a topic of interest in physics. In the context of inelastic collisions, it is observed that momentum is conserved, whereas kinetic energy is not conserved. During an inelastic collision, the involved objects have the potential to either adhere to one another or experience deformation, leading to a dissipation of kinetic energy. This wasted energy is subsequently transformed into alternative forms, such as thermal or acoustic energy. In the context of mathematics, the principle of conservation of momentum can be mathematically expressed for an inelastic collision involving two objects with masses denoted as "m1" and "m2", initial velocities denoted as "v1i" and "v2i", and final velocity denoted as "vf". This principle holds as the objects in question adhere together following the collision:

$$m1 * v1i + m2 * v2i = (m1 + m2) * vf$$

In contrast to elastic collisions, wherein the total kinetic energy remains constant before and after the collision, inelastic collisions decrease the final kinetic energy compared to the initial kinetic energy (Vittmann et al., 2022).

1.8.4. Recoil and Rocket Propulsion

The principles of conservation of momentum are effectively demonstrated through recoil and rocket propulsion. Recoil refers to the retrograde motion experienced by an object due to the forward momentum acquired by a separate object. One illustrative instance pertains to the recoil exhibited by firearms upon discharge. Upon expulsion from the gun barrel, the bullet undergoes a transfer of momentum in the forward direction, which can be attributed to the propulsive force generated by the expansion of gases resulting from the combustion of gunpowder. Due to the momentum conservation principle, the firearm undergoes an equivalent yet opposing momentum in the rearward direction, resulting in its recoil. The magnitude of the recoil force encountered by the firearm is directly proportional to the bullet's mass and velocity (Figure 1.15) (Winterberg, 2012).

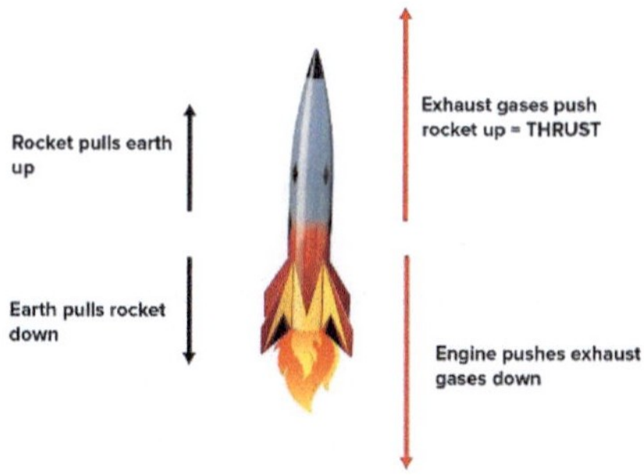

Figure 1.15. Illustration of Rocket propulsion (Source: Byju's, Creative Commons License)

Rocket propulsion is derived from the fundamental principle of Newton's third law of motion, which posits that any action will result in an equal and opposite reaction. A rocket achieves forward propulsion through mass ejection at a high velocity in the opposite direction. The expelled mass, referred to as the exhaust, gains momentum in a specific direction, resulting in the rocket experiencing an equal and opposite momentum in the opposite direction. This phenomenon propels the rocket in a forward direction. The process is governed by the principle of conservation of momentum, which enables the rocket to generate thrust and propel itself in the vacuum of space (Elsayed, 2012).

1.8.5. Momentum Applications in Sports and Transportation

Momentum has diverse applications in sports and transportation, enhancing performance, safety, and efficiency in both fields.

In Sports:

1. Baseball and Softball: In baseball and softball, players use momentum to increase the power of their swings and throws. Batters generate more force by transferring their weight from their back to their front foot during the swing, increasing the bat's momentum. Pitchers also utilize their lower body and arm movements to build momentum, leading to faster pitches (Salim & bin Aminuddin, 2022).
2. Track and Field: Momentum is vital in events like long jump, pole vault, and shot put. Athletes build momentum during their approach run, allowing them to achieve greater distances or heights when executing their jumps or throws.

3. Football: Players exploit momentum to gain an advantage over opponents. Running backs build momentum before impact to gain extra yardage through powerful collisions. On the defensive side, players utilize momentum to tackle opponents effectively (Datta & Sen, 2013).

In Transportation:

1. Automotive Safety: The concept of momentum is central to automotive safety design. Crumple zones in modern vehicles are engineered to absorb and dissipate the kinetic energy during collisions, increasing the time of impact and reducing the force experienced by occupants. Airbags also deploy based on the deceleration rate, further leveraging momentum to protect passengers.
2. Cycling and Skateboarding: Momentum is crucial in cycling and skateboarding to maintain speed and balance. Riders use momentum by shifting their weight and adjusting their posture during turns and maneuvers (Adare et al., 2008).
3. Public Transportation: In large-scale transportation systems like trains and buses, understanding momentum is essential for efficient braking and acceleration. Proper momentum control minimizes energy consumption and ensures smooth and comfortable rides for passengers (Nazeer et al., 2023).

The comprehension and efficient utilization of momentum are crucial in enhancing performance, ensuring safety, and maximizing energy efficiency in sports and transportation domains. By capitalizing on the principles of momentum, athletes can accomplish extraordinary achievements in sports, while transportation systems can be engineered to offer enhanced experiences and optimize overall performance (Li et al., 2004).

1.9. SPECIAL TOPICS IN CLASSICAL MECHANICS

Special Topics in Classical Mechanics explore advanced concepts and applications beyond traditional fundamentals. Some of these topics include:

1. Relativistic Mechanics: The applicability of classical mechanics, grounded in Newtonian physics, is limited to scenarios involving low velocities. However, as velocities approach the speed of light, the influence of relativistic effects becomes increasingly pronounced. The special relativity theory proposed by Albert

Einstein introduces the concept of relativistic mechanics, which expands upon classical mechanics by incorporating relativistic phenomena, including time dilation and length contraction (Marshman & Singh, 2015).

2. Lagrangian and Hamiltonian Mechanics: Lagrangian and Hamiltonian mechanics offer alternative formalisms for describing the dynamics of systems. They provide a more elegant and powerful way to derive equations of motion, particularly for complex systems with constraints and multiple degrees of freedom. These formulations are commonly used in analytical mechanics and quantum mechanics.
3. Celestial Mechanics: Celestial mechanics is concerned with the motion of celestial bodies, such as planets, moons, asteroids, and comets, under the influence of gravitational forces. It involves studying orbital dynamics, gravitational interactions, and the long-term stability of celestial systems (Chevrier et al., 2013).
4. Chaos and Nonlinear Dynamics: Classical mechanics can exhibit chaotic behavior in certain systems, where small changes in initial conditions lead to dramatically different outcomes. Chaos theory and nonlinear dynamics explore the behavior of complex systems, uncovering fascinating phenomena like fractals and strange attractors.
5. Rigid Body Dynamics: Rigid body dynamics deals with the motion of objects that maintain their shape during motion. This topic becomes essential when studying systems like rotating objects, gyroscopes, and celestial bodies with non-spherical shapes (Laskin, 2013).

These special topics expand classical mechanics' scope, revealing the fascinating complexities of physical phenomena and providing essential insights for understanding various systems, from subatomic particles to celestial bodies (Golmankhaneh et al., 2023).

1.9.1. Rotational Motion and Torque

The concepts of rotational motion and torque hold significant importance in classical mechanics, specifically moving objects around a stationary axis. Linear motion pertains to the movement of objects along a straight path, whereas rotational motion encompasses the spinning or rotation of objects around an axis. In contrast, torque can be understood as the rotational counterpart to force, as it induces angular acceleration in objects (Duman et al., 2015).

Rotational motion refers to the physical phenomenon where an object undergoes circular motion around a stationary axis, as opposed to linear motion along a straight path. Rotational motion describes moving objects such as wheels, propellers, and planets. In contrast to linear motion, rotational motion encompasses angular quantities such as angular displacement, angular velocity, and angular acceleration (Bhatnagar & Bhardwaj, 1994).

The angular displacement (θ) quantifies the magnitude of the rotation angle experienced by an object, usually expressed in radians. The angular velocity (ω) measures the rate at which angular displacement changes over time, and it is quantified in units of radians per second. Angular acceleration, denoted as α, refers to the derivative of angular velocity concerning time. It is quantified in units of radians per second squared.

The rotational motion equations resemble those governing linear motion, as they involve angular counterparts of displacement, velocity, and acceleration that are interconnected through time derivatives (Liu et al., 2020). Torque: Torque can be understood as the rotational counterpart to force in linear motion. Angular momentum quantifies the rotational force exerted on an object, resulting in its rotation around a specific axis. The term "torque" (τ) is mathematically expressed as the multiplication of the force (F) exerted on an object and the perpendicular distance (r) between the axis of rotation and the line along which the force acts. Mathematically, torque is expressed as $\tau = r * F * \sin(\theta)$, where θ is the angle between the force vector and the position vector (Figure 1.16) (Kane, 1970).

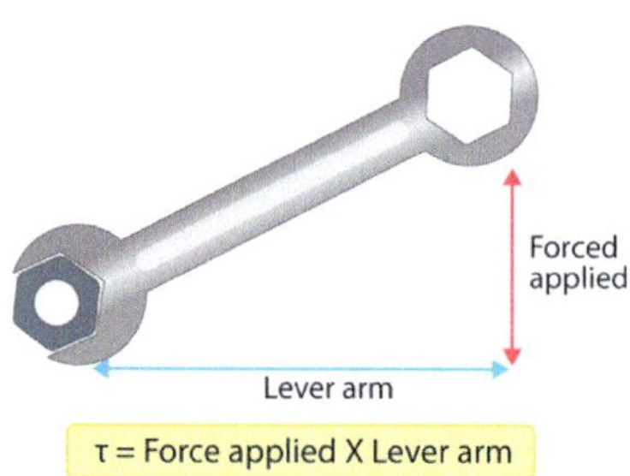

Figure 1.16. Schematic of the torque (Source: Byju's, Creative Commons License)

Torque causes an object to undergo angular acceleration, and the relationship between torque, moment of inertia (a measure of an object's resistance to rotation), and angular acceleration is given by Newton's second law for rotational motion: $\tau = I * \alpha$, where I is the moment of inertia of the object. Torque is critical in various real-world applications, such as operating motors, engines, and machinery. Engineers and designers must consider torque when designing mechanical systems to ensure proper functioning and stability (Lipfert et al., 2015).

1.9.2. Angular Momentum and its Conservation

Angular momentum is a fundamental concept within classical mechanics, serving as a crucial measure of the rotational movement exhibited by an object as it revolves around a stationary axis. The concept in question resembles linear momentum, albeit specifically about rotational motion. The concept of angular momentum (L) can be mathematically expressed as the multiplication of an object's moment of inertia (I) and its angular velocity (ω) (Simmons & Burridge, 1981).

The equation for angular momentum (L) can be expressed as the product of the moment of inertia (I) and angular velocity (ω).

The moment of inertia quantifies an object's ability to resist alterations in its rotational motion. The rotational behavior of an object is contingent upon its mass distribution and the specific axis of rotation. The moment of inertia for a point mass undergoing rotational motion about a stationary axis can be expressed as $I = m * r^2$, where "m" represents the mass of the object and "r" denotes the distance from the axis of rotation (Wolfenstein, 1951).

The conservation of angular momentum is considered to be a fundamental principle within the field of classical mechanics. Based on this principle, the aggregate angular momentum of an isolated system remains constant in the absence of any external torques acting upon it. But, when no external forces influence rotational motion, the sum of angular momentum before an occurrence is equivalent to the sum of angular momentum after the occurrence (Courtial et al., 1997). The principle of conservation of angular momentum can be mathematically represented as follows:

The conservation of angular momentum principle states that the initial angular momentum of a system is equal to the final angular momentum of the system (Kim & Gbur, 2012).

Angular momentum conservation is crucial in various physical phenomena and has practical applications in engineering and astrophysics (Becattini et al., 2008):

1. Celestial Motion: Angular momentum conservation plays a significant role in the motion of celestial bodies. Planets, moons, and stars maintain their orbits due to the conservation of angular momentum between gravitational forces and their rotational motion.
2. Gyroscopic Stability: Gyroscopes utilize angular momentum conservation to maintain stability. They are widely used in navigation systems, stabilizers in ships and airplanes, and various engineering applications.
3. Angular Momentum Transfer: In various engineering systems like flywheels and turbine engines, angular momentum is transferred between rotating components to store and release energy efficiently (Gregg et al., 2015).
4. Conservation in Quantum Mechanics: The conservation of angular momentum is extended into quantum mechanics, which plays a fundamental role in understanding atomic and subatomic behavior (Thomas & Marchiano, 2003).

The concept of angular momentum holds significant importance within classical mechanics as it pertains to the rotational motion of objects and the principle of conservation of rotational quantities. The principle of angular momentum conservation offers significant insights into a broad spectrum of physical phenomena, from celestial motion to engineering applications. Comprehending angular momentum conservation enables scientists and engineers to scrutinize and forecast the dynamics of rotating systems and devise inventive technologies that exploit the fundamental principles of classical mechanics (Gui et al., 2021).

1.9.3. Simple Harmonic Motion and Pendulums

Simple Harmonic Motion (SHM) is a fundamental form of periodic motion characterized by the oscillation of an object around an equilibrium position. This motion is governed by a restoring force directly proportional to the object's displacement from its equilibrium position and acts in the opposite direction. Simple Harmonic Motion (SHM) is a frequently observed phenomenon in many physical systems, encompassing springs, pendulums, vibrating strings, and even the orbital motion of celestial bodies (Figure 1.17) (Afandi, 2018).

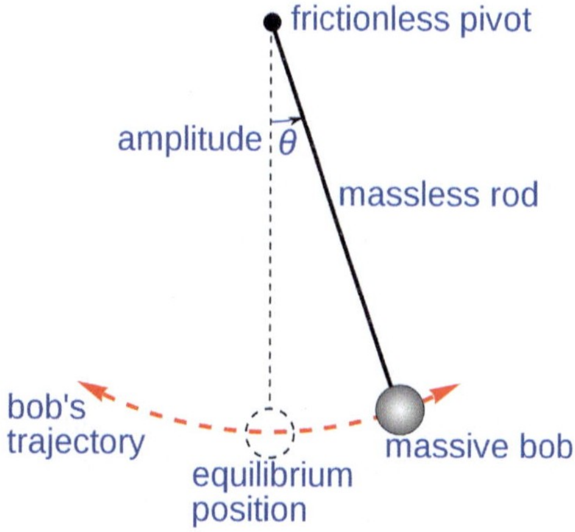

Figure 1.17. Simple gravity pendulum (Source: Chetvorno, Creative Commons License)

Pendulums exemplify the fundamental principles of simple harmonic motion. A pendulum comprises a mass, commonly called the bob, affixed to a stationary point through a string or a rod. When the pendulum is perturbed from its state of equilibrium and subsequently released, it undergoes oscillatory motion characterized by a consistent pattern of back-and-forth movement. The analysis of the motion of a simple pendulum can be conducted by employing the equations of Simple Harmonic Motion (SHM). The pendulum experiences a restoring force, which is attributed to the gravitational force and is directly proportional to the displacement of the pendulum bob from its equilibrium position. The period (T) of a pendulum, denoting the duration required to complete a single oscillation, is contingent upon the length of the pendulum (L) and the acceleration due to gravity (g). This relationship can be mathematically determined using the following formula (Richardson & Brittle, 2012):

$$\text{Period } (T) = 2\pi * \sqrt{(L / g)}$$

The amplitude of the pendulum's motion is the maximum displacement from the equilibrium position. As long as the amplitude is small, the pendulum's motion can be approximated as simple harmonic, and its period remains constant. This property makes pendulums valuable timekeeping devices, as their period is independent of the amplitude and can be used to construct accurate clocks (Ab. Aziz & Ab. Aziz, 2022).

Pendulums have practical applications in various fields, including timekeeping, engineering, and physics:

1. Clocks: The principle of the pendulum's regular and constant period makes it suitable for constructing accurate clocks. Before the invention of electronic timekeeping, pendulum clocks were widely used for precise timekeeping (Iradat & Alatas, 2017).
2. Engineering: Pendulums are used in various engineering applications, such as seismic dampers in buildings to reduce the effects of earthquakes and stabilizers in vehicles to maintain balance and reduce vibrations.
3. Physics: Pendulums are used in educational settings to demonstrate simple harmonic motion principles and to measure the acceleration due to gravity (Parnafes, 2010).

Overall, simple harmonic motion and pendulums provide valuable insights into the behavior of oscillating systems in classical mechanics. They are essential models for understanding periodic motion and are applied in various practical and scientific contexts (Fulcher & Davis, 1976).

1.9.4. Fluid Mechanics Basics

Fluid mechanics is the study of fluids and their behavior under various conditions. Fluids can be either liquids or gases, and fluid mechanics deals with how they flow, interact with solid surfaces, and respond to forces and pressure changes. Understanding the basics of fluid mechanics is crucial in various fields, including engineering, meteorology, oceanography, and aviation (Petrila & Trif, 2004).

1. Properties of Fluids: Fluids have unique properties that distinguish them from solids. Unlike solids, fluids can flow and take the shape of their containers. They exert pressure in all directions, influencing their behavior by density, viscosity (resistance to flow), and compressibility.
2. Fluid Statics: Fluid statics is concerned with the behavior of fluids at rest or in equilibrium. It involves the study of pressure distribution within a fluid and how pressure changes with depth (Lomax et al., 2002).
3. Fluid Dynamics: Fluid dynamics deals with the motion of fluids and how they interact with their surroundings. It includes the study of fluid flow patterns, forces acting on fluids, and the

effects of viscosity and turbulence. The Navier-Stokes equations are fundamental in fluid dynamics that describe the motion of fluids and are essential in solving complex fluid flow problems (Antontsev et al., 2002).

4. Bernoulli's Principle: The principle of Bernoulli is a fundamental concept within the field of fluid mechanics, which elucidates the correlation between the velocity of a fluid and the pressure it exerts. According to the principles of fluid dynamics, it is observed that in the context of an ideal fluid flow characterized by non-viscosity, incompressibility, and steadiness, there exists a reciprocal relationship between the fluid's velocity and the pressure it exerts. Specifically, an augmentation in the fluid speed leads to a concomitant reduction in the pressure. Conversely, a decrease in the fluid's velocity corresponds to increased pressure exerted. The utilization of Bernoulli's principle is frequently employed in the elucidation of lift production in the field of aerodynamics, as well as in the comprehension of the functioning of diverse fluid flow apparatuses such as carburetors and airplane wings.
5. Applications: Fluid mechanics find many practical applications in engineering and other fields. It is used in designing hydraulic systems, pumps, turbines, and pipelines. In aerodynamics, fluid mechanics is applied to understand and optimize the performance of aircraft and other flying objects. It is also employed in meteorology to study weather patterns and in oceanography to analyze ocean currents and wave behavior (Finnemore & Franzini, 2002).
6. Fluid Forces: When objects are submerged in fluids, they experience a range of forces. The buoyant force is an upward force encountered by an object when it is immersed in a fluid, and it is equivalent to the weight of the fluid that has been displaced. The drag force is a phenomenon that arises when an object traverses a fluid medium, characterized by the resistance encountered by the object. This resistance is contingent upon various factors, including the object's shape and velocity, as well as the viscosity of the fluid.

Fluid mechanics is an intriguing field in physics and engineering, which delves into the intricate dynamics of fluids and their intricate interactions with the surrounding environment. The investigation of fluid mechanics is imperative in comprehending and forecasting diverse phenomena in the natural world, as well as in devising efficient and secure engineering systems. Fluid mechanics encompasses many phenomena, from fluid dynamics to fluid statics. Bernoulli's principle and fluid forces play pivotal roles in this field, offering a robust framework for investigating and resolving practical challenges associated with fluid behavior (Durst & Arnold, 2008).

1.9.5. Vibrations and Waves

The fundamental principles of physics and engineering encompass the concepts of vibrations and waves, which elucidate the cyclical movement of objects and the transmission of disturbances within a given medium. These phenomena are commonly observed in various natural and engineered systems, encompassing musical instruments, seismic waves, electromagnetic waves, and sound propagation.

Vibrations encompass an entity's cyclic or oscillatory movement near a state of equilibrium. When an object is displaced from its state of equilibrium, it undergoes a restoring force that acts to return it to its original equilibrium position (Sari et al., 2021). The object subsequently surpasses the state of equilibrium due to inertia, resulting in a recurrent oscillatory movement. Vibrations can be categorized into two primary classifications: free vibrations and forced vibrations. Free vibrations are observed when an object is initiated into motion and subsequently allowed to oscillate without any external forces acting upon it. The frequency of autonomous oscillations is contingent exclusively upon the intrinsic characteristics of the object, including its mass and stiffness. In contrast, forced vibrations are generated by applying an external force or periodic input. The resonance phenomenon occurs when the frequency of forced vibrations aligns with the frequency of the driving force (Zhao et al., 2022).

Waves are disturbances that propagate through a medium, transmitting energy without any significant transfer of matter. There are two primary classifications for waves: mechanical waves and electromagnetic waves. Mechanical waves necessitate the presence of a medium, which can be in the form of a solid, liquid, or gas, to propagate. Mechanical waves encompass sound waves, water waves, and seismic waves as instances. In the context of mechanical waves, the particles within the medium undergo oscillatory motion, wherein they move back and forth, thereby facilitating the transmission of the disturbance to adjacent particles (Kapania & Raciti, 1989).

In contrast, electromagnetic waves possess the unique characteristic of not necessitating a medium for propagation, enabling them to traverse through a vacuum. The waves above comprise oscillating electric and magnetic fields that autonomously propagate through space at the velocity of light. Various types of electromagnetic waves can be observed in nature, such as visible light, radio waves, microwaves, and X-rays (Eltaher et al., 2016).

The characteristics of waves refer to the properties and attributes that define their behavior and nature. These features include amplitude, wavelength, frequency, speed, and direction of propagation. Understanding Waves possess several significant attributes: amplitude, wavelength, frequency, and wave speed.

The amplitude of a wave is a measure of the maximum displacement of particles from their equilibrium position. The wavelength is the interval between two successive points in a wave exhibiting identical phases, such as the distance from crest to crest or trough to trough (Afifah & Ellianawati, 2019). Frequency refers to the number of wave cycles traversing a specific location within a designated timeframe, and it exhibits an inverse relationship with the wave period.

The speed of a wave refers to the velocity at which it travels through a given medium and is influenced by the characteristics of that medium. In a general sense, it can be observed that waves characterized by higher frequencies exhibit correspondingly shorter wavelengths and greater wave velocities (Nelson et al., 1971).

Vibrations and waves are fundamental principles in physics and engineering, elucidating objects' periodic motion and the transmission of disturbances within a medium. Vibrations are distinguished by their repetitive oscillatory movement, whereas waves are defined as energy propagation without matter's physical displacement.

A comprehensive comprehension of vibrations and waves is imperative for a diverse array of practical applications, encompassing the design of musical instruments, the examination of seismic activity, the investigation of electromagnetic radiation, and the analysis of sound propagation and communication systems (Mahfudz & Billah, 2020).

Activity 1.1

Imagine you are designing a roller coaster. How would you apply the principles of kinematics and dynamics to create thrilling and safe rides for passengers? Consider factors like speed, acceleration, and forces experienced during the ride.

SUMMARY

This chapter presents a comprehensive examination of the fundamental concepts and principles that underpin classical mechanics. The chapter commences by defining mechanics as the scientific inquiry into the phenomenon of motion while also presenting the differentiation between linear and rotational motion. This statement underscores the significance of classical mechanics as a foundational discipline within the field of physics, serving as a basis for comprehending diverse natural phenomena and engineering implementations. The historical survey examines the significant contributions made by notable figures such as Aristotle, Galileo, and Newton to advance classical mechanics. The discourse examines Aristotle's theory of motion, Galileo's contributions to understanding inertia, and Newton's formulation of the three laws of motion. These discussions serve to illustrate the progressive development of classical mechanics throughout history. The subsequent chapter examines the breadth and practical implementations of classical mechanics across diverse domains, encompassing celestial mechanics and planetary motion, as well as commonplace situations such as friction, circular motion, and tension in ropes. The statement highlights the significance of classical mechanics as a foundational framework for the examination of real-world phenomena and the design of engineering systems. The comprehensive examination of kinematics entails introducing the fundamental notion of motion description, which involves distinguishing between scalars and vectors and elucidating the concepts of position, displacement, distance, speed, and velocity. The dynamics section encompasses the study of forces and motion, elucidating Newton's three laws of motion and exploring the fundamental forces present in nature. The chapter concludes by thoroughly examining the concepts of energy and work, encompassing the definition of energy, kinetic energy, potential energy, and the work-energy theorem. This text elucidates the principle of conservation of mechanical energy and introduces power as a metric denoting the rate at which work is performed. The chapter elucidates diverse instances where classical mechanics find practical applications, including but not limited to sports, transportation, and everyday activities. The chapter emphasizes the significance of classical mechanics as a fundamental discipline that serves as a basis for comprehending various facets of the physical realm. Additionally, it highlights its practical utility in diverse scientific and engineering fields.

REVIEW QUESTIONS

1. What is the historical significance of Galileo's contributions to mechanics, and how did they challenge Aristotle's theory of motion?
2. Define scalars and vectors, and provide examples of each in the context of classical mechanics.
3. Explain the difference between speed and velocity, and provide real-life examples to illustrate the concepts.
4. Describe the concepts of displacement and distance, and discuss how they are related but different in terms of motion analysis.
5. According to Newton's laws of motion, how does an object's motion change when acted upon by external forces?
6. Provide practical examples of applying Newton's laws in everyday scenarios, such as driving a car or playing sports.
7. Define kinetic energy and potential energy, and explain the relationship between work and energy in mechanical systems.
8. Discuss the principle of conservation of mechanical energy and provide examples of situations where it applies.
9. How does impulse affect an object's momentum, and what is the impulse-momentum theorem?
10. Describe the concept of angular momentum, and explain how it is conserved in certain situations.
11. What is simple harmonic motion, and how is it related to pendulums and oscillatory motion?
12. Briefly explain the basics of fluid mechanics and the principles governing the behavior of fluids.

REFERENCES

1. Ab. Aziz, N. A., & Ab. Aziz, K. (2022). Pendulum Search Algorithm: An Optimization Algorithm Based on Simple Harmonic Motion and Its Application for a Vaccine Distribution Problem. *Algorithms*, *15*(6), 214.
2. Adams, J., Adler, C., Aggarwal, M. M., Ahammed, Z., Amonett, J., Anderson, B. D., ... & Landgraf, J. M. (2003). Transverse-momentum and collision-energy dependence of High-p T Hadron Suppression in A u+ A u Collisions at ultrarelativistic Energies. *Physical Review Letters*, *91*(17), 172302.
3. Adare, A., Afanasiev, S., Aidala, C., Ajitanand, N. N., Akiba, Y., Al-Bataineh, H., ... & Hornback, D. (2008). Quantitative constraints on the transport properties of hot partonic matter from semi-inclusive single high transverse momentum pion suppression in Au+ Au collisions at s NN= 200 GeV. *Physical Review C*, *77*(6), 064907.
4. Afandi, Z. (2018, April). Development of gravity acceleration measurement using simple harmonic motion pendulum method based on digital technology and photogate sensor. In *IOP Conference Series: Materials Science and Engineering*. IOP Publishing. (Vol. 335, pp. 12064).
5. Afifah, R., & Ellianawati, E. (2019). Student Cognitive Profile with STEM Based Teaching Material on the Subject of Vibrations and Waves. *Jurnal Penelitian & Pengembangan Pendidikan Fisika*, *5*(2), 217–226.
6. Andal, C. K., & Jayapal, R. (2022). Design and implementation of IoT based intelligent energy management controller for PV/wind/battery system with cost minimization. *Renewable Energy Focus*, *43*, 255–262.
7. Antoniadis, I., Baessler, S., Büchner, M., Fedorov, V. V., Hoedl, S., Lambrecht, A., ... & Sobolev, Y. (2011). Short-range fundamental forces. *Comptes Rendus Physique*, *12*(8), 755–778.
8. Antontsev, S. N., Díaz, J. I., Shmarev, S., & Kassab, A. J. (2002). Energy Methods for Free Boundary Problems: Applications to Nonlinear PDEs and Fluid Mechanics. Progress in Nonlinear Differential Equations and Their Applications, Vol 48. *Appl. Mech. Rev.*, *55*(4), B74-B75.
9. Atilola, O., Valentine, S., Kim, H. H., Turner, D., McTigue, E., Hammond, T., & Linsey, J. (2014). Mechanix: A natural sketch

interface tool for teaching truss analysis and free-body diagrams. *AI EDAM*, *28*(2), 169–192.

10. Balaban, O. (1995). The modern misunderstanding of Aristotle's theory of motion. *Journal for general philosophy of science*, *26*(1), 1–10.
11. Barwell, R. (2014). Centripetal and centrifugal language forces in one elementary school second language mathematics classroom. *ZDM*, *46*, 911–922.
12. Becattini, F., Piccinini, F., & Rizzo, J. (2008). Angular momentum conservation in heavy ion collisions at very high energy. *Physical Review C*, *77*(2), 024906.
13. Berry, M. V. (1977). Semi-classical mechanics in phase space: a study of Wigner's function. *Philosophical Transactions of the Royal Society of London. Series A, Mathematical and Physical Sciences*, *287*(1343), 237–271.
14. Bhatnagar, K. B., & Bhardwaj, R. (1994). Rotational motion of a satellite in an elliptical orbit under the influence of third body torque (I). *Bulletin of the Astronomical Society of India V. 22, P. 359–367, 1994*, *22*, 359–367.
15. Blackman, R. B., & Tukey, J. W. (1958). The measurement of power spectra from the point of view of communications engineering—Part I. *Bell System Technical Journal*, *37*(1), 185–282.
16. Blackwell, R. J. (1966). Descartes' laws of motion. *Isis*, *57*(2), 220–234.
17. Bonanno, A., Bozzo, G., Grandinetti, M., & Sapia, P. (2016). Work–energy theorem and friction forces: two experiments. *Physics Education*, *51*(6), 065004.
18. Botzer, G., & Yerushalmy, M. (2008). Embodied semiotic activities and their role in the construction of mathematical meaning of motion graphs. *International Journal of Computers for Mathematical Learning*, *13*, 111–134.
19. Bray, J. C., & Eldridge, J. J. (2018). Neutron star kicks–II. Revision and further testing of the conservation of momentum 'kick'model. *Monthly Notices of the Royal Astronomical Society*, *480*(4), 5657–5672.
20. Cape, E. G., Skoufis, E. G., Weyman, A. E., Yoganathan, A. P., & Levine, R. A. (1989). A new method for noninvasive quantification of valvular regurgitation based on conservation of momentum. In vitro validation. *Circulation*, *79*(6), 1343–1353.

21. Carling, J., Williams, T. L., & Bowtell, G. (1998). Self-propelled anguilliform swimming: simultaneous solution of the two-dimensional Navier–Stokes equations and Newton's laws of motion. *Journal of experimental biology*, *201*(23), 3143–3166.
22. Chaves, D. B. (2012). Eureka! A momentum strategy that also works in Japan. *A Momentum Strategy that Also Works in Japan (January 9, 2012)*.
23. Chernetsov, N., Kishkinev, D., & Mouritsen, H. (2008). A long-distance avian migrant compensates for longitudinal displacement during spring migration. *Current Biology*, *18*(3), 188–190.
24. Chevrier, J., Madani, L., Ledenmat, S., & Bsiesy, A. (2013). Teaching classical mechanics using smartphones. *The Physics Teacher*, *51*(6), 376–377.
25. Cook, G. A., Pandit, N. R., Beaverstock, J. V., Taylor, P. J., & Pain, K. (2007). The role of location in knowledge creation and diffusion: Evidence of centripetal and centrifugal forces in the City of London financial services agglomeration. *Environment and Planning A*, *39*(6), 1325–1345.
26. Courtial, J., Dholakia, K., Allen, L., & Padgett, M. J. (1997). Second-harmonic generation and the conservation of orbital angular momentum with high-order Laguerre-Gaussian modes. *Physical Review A*, *56*(5), 4193.
27. Cunningham, B. A. (2013, March). Electromagnetism, Optics, Acoustics, Heat Transfer, Classical Mechanics, and Fluid Dynamics. In *AIP Conference Proceedings* (Vol. 1517, No. 1, pp. 208–213). American Institute of Physics.
28. Czarnocka, M., & Żytkow, J. M. (1982). Difficulties with the reduction of classical to relativistic mechanics. In *Polish Essays in the Philosophy of the Natural Sciences*. Dordrecht: Springer Netherlands. (Vol. 3, pp. 319–332).
29. Datta, A., & Sen, S. (2013). Biomedical application of energy and momentum transport in microfluidics in the presence of shear-driven flows. *Radiation Effects and Defects in Solids*, *168*(10), 852–857.
30. Del Giudice, V., De Paola, P., & Forte, F. (2016). The appraisal of office towers in bilateral monopoly's market: evidence from application of Newton's physical laws to the directional centre of Naples. *International Journal of Applied Engineering Research*, *11*(18), 9455–9459.

31. Denny, Y. R., Utami, I. S., Rohanah, S., & Muliyati, D. (2020). The development of blended learning model using Edmodo to train student critical thinking skills on impulse-momentum topic. *Jurnal Penelitian & Pengembangan Pendidikan Fisika*, *6*(1), 113–120.
32. Di Mauro, M., & Naddeo, A. (2022). Intertheoretic relations, singular limits and emergence: a critical overview of the relation between classical and quantum mechanics. *Intertheoretic relations, singular limits and emergence: a critical overview of the relation between classical and quantum mechanics*, 45, 139–144.
33. Duijzer, C., Van den Heuvel-Panhuizen, M., Veldhuis, M., & Doorman, M. (2019). Supporting primary school students' reasoning about motion graphs through physical experiences. *Zdm*, *51*, 899–913.
34. Duman, I., Demirci, N., & Sekercioglu, A. (2015). University students' difficulties and misconceptions on rolling, rotational motion and torque concepts. *International Journal on New Trends in Education and Their Implications*, *6*(1), 46–54.
35. Durst, F., & Arnold, I. (2008). *Fluid mechanics: an introduction to the theory of fluid flows*. Berlin: Springer. (Vol. 675, pp. 987–988).
36. Elsayed, T. A. (2012). Interpreting recoil for undergraduate students. *The Physics Teacher*, *50*(4), 209–211.
37. Eltaher, M. A., Khater, M. E., & Emam, S. A. (2016). A review on nonlocal elastic models for bending, buckling, vibrations, and wave propagation of nanoscale beams. *Applied Mathematical Modelling*, *40*(5–6), 4109–4128.
38. Fahrunnisa, S. A., Rismawati, Y., Sinaga, P., & Rusdiana, D. (2021, March). Experiments of the law of conservation of mechanical energy using video tracker in high school learning. In *Journal of Physics: Conference Series*. IOP Publishing. (Vol. 1806, pp. 120–135).
39. Field, G. B., Chaisson, E. J., Field, G. B., & Chaisson, E. J. (1985). Forces of Nature: The Limits of Knowledge. *The Invisible Universe: Probing the frontiers of astrophysics*, 139–151.
40. Finnemore, E. J., & Franzini, J. B. (2002). *Fluid mechanics with engineering applications*. McGraw-Hill Education. 78(5), 98522.
41. Firdawati, R., Maison, M., & Nazarudin, N. (2021). Development of Mobile Learning Media on Newton's Laws Using the Appy Pie Application. *Jurnal Penelitian Pendidikan IPA*, *7*(2), 202–206.

42. Fulcher, L. P., & Davis, B. F. (1976). Theoretical and experimental study of the motion of the simple pendulum. *American Journal of Physics*, *44*(1), 51–55.

43. Gartland, P. O., & Slagsvold, B. J. (1975). Transitions conserving parallel momentum in photoemission from the (111) face of copper. *Physical Review B*, *12*(10), 4047.

44. Gateno, J., Anderson, P. B., Xia, J. J., Horng, J. C., Teichgraeber, J. F., & Liebschner, M. A. (2004). A comparative assessment of mandibular condylar position in patients with anterior disc displacement of the temporomandibular joint. *Journal of Oral and Maxillofacial surgery*, *62*(1), 39–43.

45. Ginchev, I., Guerraggio, A., & Rocca, M. (2006). From scalar to vector optimization. *Applications of Mathematics*, *51*(1), 5–36.

46. Golmankhaneh, A. K., Welch, K., Tunç, C., & Gasimov, Y. S. (2023). Classical mechanics on fractal curves. *The European Physical Journal Special Topics*, 8(4), 1–9.

47. Gough, W. (1990). Mixing scalars and vectors-an elegant view of physics. *European Journal of Physics*, *11*(6), 326.

48. Graham, D. W. (1988). Aristotle's Definition of Motion. *Ancient Philosophy*, *8*(2), 209–215.

49. Gregg, P., Kristensen, P., & Ramachandran, S. (2015). Conservation of orbital angular momentum in air-core optical fibers. *Optica*, *2*(3), 267–270.

50. Gui, G., Brooks, N. J., Kapteyn, H. C., Murnane, M. M., & Liao, C. T. (2021). Second-harmonic generation and the conservation of spatiotemporal orbital angular momentum of light. *Nature Photonics*, *15*(8), 608–613.

51. Halilovic, A., Mešić, V., Hasović, E., & Vidak, A. (2022). The Post-Instruction Conceptions about Conservation of Mechanical Energy: Findings from a Survey research with High School and University Students. *Journal of Turkish Science Education*, *19*(1), 144–162.

52. Halloun, I. (1998). Schematic concepts for schematic models of the real world: The Newtonian concept of force. *Science Education*, *82*(2), 239–263.

53. Havas, P. (1964). Four-dimensional formulations of Newtonian mechanics and their relation to the special and the general theory of relativity. *Reviews of Modern Physics*, *36*(4), 938.

54. Henneaux, M., & Teitelboim, C. (1988). BRST cohomology in classical mechanics. *Communications in mathematical physics*, *115*, 213–230.
55. Husna, A., & Fajar, D. M. (2022). Development of interactive learning media based on articulate storyline 3 on Newton's law material with a contextual approach at the junior high school level. *IJIS Edu: Indonesian Journal of Integrated Science Education*, *4*(1), 17–26.
56. Iradat, R. D., & Alatas, F. (2017, September). The implementation of problem-solving based laboratory activities to teach the concept of simple harmonic motion in senior high school. In *Journal of Physics: Conference Series*. IOP Publishing. (Vol. 895, pp. 120–125).
57. Izquierdo, S., Montañés, C., Dopazo, C., & Fueyo, N. (2010). Analysis of CSP plants for the definition of energy policies: The influence on electricity cost of solar multiples, capacity factors and energy storage. *Energy policy*, *38*(10), 6215–6221.
58. Jimoyiannis, A., & Komis, V. (2001). Computer simulations in physics teaching and learning: a case study on students' understanding of trajectory motion. *Computers & education*, *36*(2), 183–204.
59. Kafle, P. R., Sharma, S., Lewis, G. F., Robotham, A. S., & Driver, S. P. (2018). The need for speed: escape velocity and dynamical mass measurements of the Andromeda galaxy. *Monthly Notices of the Royal Astronomical Society*, *475*(3), 4043–4054.
60. Kane, T. R. (1970). Solution of the equations of rotational motion for a class of torque-free gyrostats. *AIAA Journal*, *8*(6), 1141–1143.
61. Kapania, R. K., & Raciti, S. (1989). Recent advances in analysis of laminated beams and plates, Part II: Vibrations and wave propagation. *AIAA journal*, *27*(7), 935–946.
62. Kawase, H., Sánchez-Sesma, F. J., & Matsushima, S. (2011). The optimal use of horizontal-to-vertical spectral ratios of earthquake motions for velocity inversions based on diffuse-field theory for plane waves. *Bulletin of the Seismological Society of America*, *101*(5), 2001–2014.
63. Kim, S. M., & Gbur, G. (2012). Angular momentum conservation in partially coherent wave fields. *Physical Review A*, *86*(4), 043814.
64. Kolmogorov, A. N., & Tichomirov, V. M. (1991). *Selected works: Mathematics and mechanics*. Kluwer Acad. Publ. (Vol. 2, pp. 252–261).

65. Konno, K., & Ohmachi, T. (1998). Ground-motion characteristics estimated from spectral ratio between horizontal and vertical components of microtremor. *Bulletin of the Seismological Society of America*, *88*(1), 228–241.
66. Kurniawan, B. R., Fawaiz, S., Ramadani, C. I., Affriyenni, Y., Darmawan, A., & Wibowo, F. C. (2023, January). What do teachers need to teach free-body diagrams effectively? A need analysis in development country. In *AIP Conference Proceedings*. AIP Publishing. (Vol. 2540, pp. 242–245).
67. Lamanepa, G. H., Maing, C. M., Mukin, M. U. J., & Naen, A. B. (2022). The Role of Visual Representation for High School Physics in Teaching of Classical Mechanics. *Jurnal Penelitian & Pengembangan Pendidikan Fisika*, *8*(1), 105–114.
68. Laskin, N. (2013). Generalized classical mechanics. *The European Physical Journal Special Topics*, *222*(8), 1929–1938.
69. Lawson, R. A., & McDermott, L. C. (1987). Student understanding of the work-energy and impulse-momentum theorems. *American Journal of Physics*, *55*(9), 811–817.
70. Li, F. C., Kawaguchi, Y., & Hishida, K. (2004). Investigation on the characteristics of turbulence transport for momentum and heat in a drag-reducing surfactant solution flow. *Physics of Fluids*, *16*(9), 3281–3295.
71. Lipfert, J., van Oene, M. M., Lee, M., Pedaci, F., & Dekker, N. H. (2015). Torque spectroscopy for the study of rotary motion in biological systems. *Chemical reviews*, *115*(3), 1449–1474.
72. Liu, Y., Hou, W., Han, X., & Zang, J. (2020). Three-dimensional dynamics of a magnetic hopfion driven by spin transfer torque. *Physical review letters*, *124*(12), 127204.
73. Lomax, H., Pulliam, T. H., Zingg, D. W., & Kowalewski, T. A. (2002). Fundamentals of computational fluid dynamics. *Appl. Mech. Rev.*, *55*(4), B61-B61.
74. Lorenz, E. N. (1955). Available potential energy and the maintenance of the general circulation. *Tellus*, *7*(2), 157–167.
75. Mahfudz, A. Z., & Billah, A. (2020, June). The development of android-based learning media on vibrations and waves topic for junior high school students. In *Journal of Physics: Conference Series*. IOP Publishing. (Vol. 1567, pp. 2009).

76. Marshman, E., & Singh, C. (2015). Framework for understanding the patterns of student difficulties in quantum mechanics. *Physical Review Special Topics-Physics Education Research*, *11*(2), 020119.

77. McDermott, L. C., Rosenquist, M. L., & Van Zee, E. H. (1987). Student difficulties in connecting graphs and physics: Examples from kinematics. *American Journal of Physics*, *55*(6), 503–513.

78. Mehar, A., Chandra, S., & Velmurugan, S. (2013). Speed and acceleration characteristics of different types of vehicles on multi-lane highways. *European Transport*, *55*(1), 1–12.

79. Meier, D. L., & Schroder, D. K. (1984). Contact resistance: Its measurement and relative importance to power loss in a solar cell. *IEEE Transactions on Electron Devices*, *31*(5), 647–653.

80. Methfessel, M., Hennig, D., & Scheffler, M. (1992). Trends of the surface relaxations, surface energies, and work functions of the 4d transition metals. *Physical Review B*, *46*(8), 4816.

81. Mintzberg, H. (1991). The effective organization: forces and forms. *MIT Sloan Management Review*, *32*(2), 54.

82. Mitiche, A., Seida, S., & Aggarwal, J. K. (1988). Using constancy of distance to estimate position and displacement in space. *IEEE Transactions on Pattern Analysis and Machine Intelligence*, *10*(4), 594–599.

83. Montoya, M. E. (1999). Silence and silencing: Their centripetal and centrifugal forces in legal communication, pedagogy and discourse. *U. Mich. JL Reform*, *33*, 263.

84. Moo, J. G. S., Mayorga-Martinez, C. C., Wang, H., Teo, W. Z., Tan, B. H., Luong, T. D., ... & Pumera, M. (2018). Bjerknes forces in motion: Long-range translational motion and chiral directionality switching in bubble-propelled micromotors via an ultrasonic pathway. *Advanced Functional Materials*, *28*(25),

85. Muckerman, J. T. (1981). Applications of classical trajectory techniques to reactive scattering. *Theor. Chem.: Adv. Perspect.;(United States)*, *6*, 510121.

86. Murphy, A. H. (1972). Scalar and vector partitions of the probability score: Part II. N-state situation. *Journal of Applied Meteorology and Climatology*, *11*(8), 1183–1192.

87. Nazeer, M., Nazir, M. W., Ali, N., Javed, T., Abdelmohsen, S. A., & Khan, M. I. (2023). Momentum and thermal transport analysis in MHD

nanofluid through the thermally heated square conduit: Finite element method. *Journal of Magnetism and Magnetic Materials*, 4, 170954.

88. Nelson, R. B., Dong, S. B., & Kalra, R. D. (1971). Vibrations and waves in laminated orthotropic circular cylinders. *Journal of Sound and Vibration*, *18*(3), 429–444.

89. Nowakowski, M., & Rosu, H. C. (2002). Newton's laws of motion in the form of a Riccati equation. *Physical Review E*, *65*(4), 047602.

90. Owens, B. P., Baker, W. E., Sumpter, D. M., & Cameron, K. S. (2016). Relational energy at work: Implications for job engagement and job performance. *Journal of Applied Psychology*, *101*(1), 35.

91. Pachava, S., Dharmavarapu, R., Vijayakumar, A., Jayakumar, S., Manthalkar, A., Dixit, A., ... & Bhattacharya, S. (2020). Generation and decomposition of scalar and vector modes carrying orbital angular momentum: a review. *Optical Engineering*, *59*(4), 041205–041205.

92. Parnafes, O. (2010). When simple harmonic motion is not that simple: Managing epistemological complexity by using computer-based representations. *Journal of Science Education and Technology*, *19*, 565–579.

93. Patel, V., Eldho, T. I., & Prabhu, S. V. (2018). Theoretical study on the prediction of the hydrodynamic performance of a Savonius turbine based on stagnation pressure and impulse momentum principle. *Energy conversion and management*, *168*, 545–563.

94. Patil, D. P., Wani, A. N., & Thete, P. D. (2022). Applications of Karry-Kalim-Adnan Transformations (KKAT) to Newtons Law of Cooling. *International Journal of Scientific Development and Research*, *7*(12), 1024–1030.

95. Patra, S. K., & Haldar, S. (2022). Seismic performance of multimegawatt offshore wind turbines in liquefiable soil under horizontal and vertical motions. *International Journal of Geomechanics*, *22*(3), 04021305.

96. Petrila, T., & Trif, D. (2004). *Basics of fluid mechanics and introduction to computational fluid dynamics.* Springer Science & Business Media. (Vol. 3, pp. 987–990).

97. Pierratos, T., & Polatoglou, H. M. (2017). Study of the conservation of mechanical energy in the motion of a pendulum using a smartphone. *Physics Education*, *53*(1), 015021.

98. Pili, U. B., & Violanda, R. R. (2021). Tracker-assisted modelling: simultaneous validation of the conservation law of energy and the work-energy theorem. *Physics Education*, *57*(1), 015012.

99. Pride, T. O. B., Vokos, S., & McDermott, L. C. (1998). The challenge of matching learning assessments to teaching goals: An example from the work-energy and impulse-momentum theorems. *American Journal of Physics*, *66*(2), 147–157.
100. Radford, L., Miranda, I., & Guzmán, J. (2008, July). Relative motion, graphs and the heteroglossic transformation of meanings: A semiotic analysis. In *Proceedings of the Joint 32nd conference of the international group for the psychology of mathematics education and the 30th North American chapter* (Vol. 4, pp. 161–168).
101. Richardson, T. H., & Brittle, S. A. (2012). Physical pendulum experiments to enhance the understanding of moments of inertia and simple harmonic motion. *Physics Education*, *47*(5), 537.
102. Rodriguez-Maldonado, J. (2019). Estimation of angular velocity and acceleration with Kalman filter, based on position measurement only. *Measurement*, *145*, 130–136.
103. SafarPour, H., & Ghadiri, M. (2017). Critical rotational speed, critical velocity of fluid flow and free vibration analysis of a spinning SWCNT conveying viscous fluid. *Microfluidics and Nanofluidics*, *21*, 1–23.
104. Saglam-Arslan, A., & Devecioglu, Y. (2010, April). Student teachers' levels of understanding and model of understanding about Newton's laws of motion. In *Asia-Pacific Forum on Science Learning & Teaching* (Vol. 11, pp. 6540–6543).
105. Sakalayen, Q. M. H., Duru, O., & Hirata, E. (2021). An econophysics approach to forecast bulk shipbuilding orderbook: an application of Newton's law of gravitation. *Maritime Business Review*, *6*(3), 234–255.
106. Salim, W. S. I. W., & bin Aminuddin, A. (2022). Aerodynamic Analysis of Propeller for Heavy Lifting Drone Applications Using Blade Element Momentum Analysis. *Progress in Aerospace and Aviation Technology*, *2*(2), 24–32.
107. Saltiel, E., & Malgrange, J. L. (1980). 'Spontaneous' ways of reasoning in elementary kinematics. *European Journal of Physics*, *1*(2), 73.
108. Sanjuán, M. A. (2021). A modern approach to teaching classical mechanics: A review of Modern classical mechanics by TM Helliwell and VV Sahakian, Cambridge, Cambridge University Press, 2020, 704 pp.,£ 56.00,(hardback), ISBN 978–1-10–883497–1. Scope: textbook, reference. Level: undergraduate, advanced undergraduate, postgraduate, teacher, scientist, engineers. *Contemporary Physics*, *62*(1), 46–49.

109. Sansón, L. Z., Aguiar, A. B., & van Heijst, G. J. F. (2012). Horizontal and vertical motions of barotropic vortices over a submarine mountain. *Journal of Fluid Mechanics*, *695*, 173–198.

110. Sari, D. K., Wea, V. M., Rahayu, M., & Dinata, P. A. C. (2021). The comic isomorphic physics series of vibrations and waves to increase the learning interest of junior high school students. *Gravity: Jurnal Ilmiah Penelitian dan Pembelajaran Fisika*, *7*(2), 6541–6542.

111. Sarlet, W., & Cantrijn, F. (1981). Generalizations of Noether's theorem in classical mechanics. *Siam Review*, *23*(4), 467–494.

112. Schlegel, H. B. (2003). Exploring potential energy surfaces for chemical reactions: an overview of some practical methods. *Journal of Computational Chemistry*, *24*(12), 1514–1527.

113. Sekihara, K., Nagarajan, S. S., Poeppel, D., & Marantz, A. (2004). Asymptotic SNR of scalar and vector minimum-variance beamformers for neuromagnetic source reconstruction. *IEEE Transactions on Biomedical Engineering*, *51*(10), 1726–1734.

114. Shekhar, S., Caglar, F., Dukeman, A., Hou, L., Gokhale, A., Kinnebrew, J. S., & Biswas, G. (2014, June). A Collaborative K-12 STEM Education Framework Using Traffic Flow as a Real-world Challenge Problem. In *2014 ASEE Annual Conference & Exposition* (Vol. 2, pp. 24–28).

115. Sidhu, T. S. (1999). Accurate measurement of power system frequency using a digital signal processing technique. *IEEE Transactions on Instrumentation and Measurement*, *48*(1), 75–81.

116. Simmons, A. J., & Burridge, D. M. (1981). An energy and angular-momentum conserving vertical finite-difference scheme and hybrid vertical coordinates. *Monthly Weather Review*, *109*(4), 758–766.

117. Sugiyama, K., & Yokoyama, K. (1996). Displacement of the endotracheal tube caused by change of head position in pediatric anesthesia: evaluation by fiberoptic bronchoscopy. *Anesthesia & Analgesia*, *82*(2), 251–253.

118. Suh, Y. S. (2019). Laser sensors for displacement, distance and position. *Sensors*, *19*(8), 1924.

119. Sun, H., Zhang, Y., Baleanu, D., Chen, W., & Chen, Y. (2018). A new collection of real world applications of fractional calculus in science and engineering. *Communications in Nonlinear Science and Numerical Simulation*, *64*, 213–231.

120. Szabados, L. B. (2009). Quasi-local energy-momentum and angular momentum in general relativity. *Living Reviews in Relativity*, *12*, 1–163.
121. Taylor, J. R., & Taylor, J. R. (2005). *Classical Mechanics*. Sausalito, CA: University Science Books. (Vol. 1, pp. 654–656).
122. Thomas, J. L., & Marchiano, R. (2003). Pseudo angular momentum and topological charge conservation for nonlinear acoustical vortices. *Physical Review Letters*, *91*(24), 244302.
123. Thornton, R. K., & Sokoloff, D. R. (1998). Assessing student learning of Newton's laws: The force and motion conceptual evaluation and the evaluation of active learning laboratory and lecture curricula. *american Journal of Physics*, *66*(4), 338–352.
124. Topping, J. (1927). On the mutual potential energy of a plane network of doublets. *Proceedings of the Royal Society of London. Series A, Containing Papers of a Mathematical and Physical Character*, *114*(766), 67–72.
125. Ültay, E., & Alev, N. (2017). Investigating the Effect of the Activities Based on Explanation Assisted REACT Strategy on Learning Impulse, Momentum and Collisions Topics. *Journal of Education and Practice*, *8*(7), 174–186.
126. Valtonen, M., Anosova, J., Kholshevnikov, K., Mylläri, A., Orlov, V., Tanikawa, K., ... & Tanikawa, K. (2016). From newton to einstein: The discovery of laws of motion and gravity. *The Three-body Problem from Pythagoras to Hawking*, 4(2), 31–49.
127. Vedral, V. (2010). Quantum Statistical Mechanics, by WC Schieve and LP Horwitz: Scope: basic theory, with applications from quantum optics to quantum gravity. *Level: Researchers and Advanced Students.6*(5), 9871–9875.
128. Vittmann, C., Kessing, R. K., Lim, J., Huelga, S. F., & Plenio, M. B. (2022). Interface-induced conservation of momentum leads to chiral-induced spin selectivity. *The Journal of Physical Chemistry Letters*, *13*(7), 1791–1796.
129. Waldrip, B., Prain, V., & Sellings, P. (2013). Explaining Newton's laws of motion: Using student reasoning through representations to develop conceptual understanding. *Instructional Science*, *41*, 165–189.
130. Wallace, W. A. (1971). Mechanics from Bradwardine to Galileo. *Journal of the History of Ideas*, *32*(1), 15–28.

131. Watkins, E. (1997). The laws of motion from Newton to Kant. *Perspectives on Science*, *5*(3), 311–348.

132. Weiss, M., Knirsch, W., Kretschmar, O., Dullenkopf, A., Tomaske, M., Balmer, C., ... & Berger, F. (2006). Tracheal tube-tip displacement in children during head-neck movement—a radiological assessment. *BJA: British Journal of Anaesthesia*, *96*(4), 486–491.

133. White, B. Y. (1984). Designing computer games to help physics students understand Newton's laws of motion. *Cognition and Instruction*, *1*(1), 69–108.

134. Winterberg, F. (2012). Matter–antimatter gigaelectron volt gamma ray laser rocket propulsion. *Acta Astronautica*, *81*(1), 34–39.

135. Wolfenstein, L. (1951). Conservation of angular momentum in the statistical theory of nuclear reactions. *Physical Review*, *82*(5), 690.

136. Xu, T., Zhang, S., Jiang, Z., Liu, Z., & Cheng, H. (2020). Collision avoidance of high-speed obstacles for mobile robots via maximum-speed aware velocity obstacle method. *IEEE Access*, *8*, 138493–138507.

137. Yaghobi, H. (2016). Out-of-step protection of generator using analysis of angular velocity and acceleration data measured from magnetic flux. *Electric Power Systems Research*, *132*, 9–21.

138. Yang, V. X., Gordon, M. L., Qi, B., Pekar, J., Lo, S., Seng-Yue, E., ... & Vitkin, I. A. (2003). High speed, wide velocity dynamic range Doppler optical coherence tomography (Part I): System design, signal processing, and performance. *Optics Express*, *11*(7), 794–809.

139. Zhao, Z., Chen, Y., Hu, X., Bao, R., Wu, B., & Chen, W. (2022). Vibrations and waves in soft dielectric elastomer structures. *International Journal of Mechanical Sciences*, 41(45), 107885.

140. Ziefle, R. G. (2016). Unification of the four fundamental forces of nature by a binary quantum model. *Physics Essays*, *29*(1), 982–985.

CHAPTER 2

Kinematics: Describing Motion

Contents

2.1. INTRODUCTION

As an essential discipline within classical mechanics, Kinematics analyzes motion while disregarding the underlying factors that drive motion, including forces and energy. The primary objective of this study is to provide a comprehensive description of the position, acceleration, and displacement of objects as they undergo motion in both spatial and temporal dimensions. Kinematics offers the necessary terminology and methodologies to comprehend and evaluate the movement of entities quantitatively. This field allows us to analyze and forecast their actions without looking into the underlying forces that drive their motion. Kinematics employs mathematical principles to model and elucidate the characteristics of motion, rendering it an essential component within engineering and physics. The initial stage of examining the movement of objects serves as the foundational basis for comprehending intricate principles in mechanics and various domains within the field of physics (Saltiel & Malgrange, 1980).

The investigation of motion holds significant importance due to its fundamental role in advancing the comprehension of the natural world and its critical contributions to diverse scientific and engineering domains. A fundamental component of the study of Kinematics involves comprehending the concept of an object's position, which pertains to its spatial coordinates at a specific moment in time. The concept of position is fundamental in the study of motion, as it can be accurately described using a coordinate system. The concept of displacement arises from alterations in position, specifically referring to the disparity between an object's final and initial positions. Displacement is a vector quantity possessing both magnitude and direction, enabling the characterization of an object's motion direction and the distance it has traversed from its initial position. Velocity is classified as a vector quantity due to its incorporation of speed and direction of motion (Waldron & Schmiedeler, 2016).

An instance can be observed in which an object, while in motion along a linear path at a consistent pace, will possess an unchanging velocity aligned with the direction of its movement. Acceleration is defined as the derivative of velocity concerning time. When the velocity of an object changes, it is subject to acceleration. Acceleration can arise due to the presence of external forces exerted on an object, such as the force of gravity or the force of friction. An illustrative example is an object descending towards the Earth, which undergoes an acceleration resulting from the force of gravity. This acceleration leads to a progressive augmentation of the object's velocity over a given period (Chen et al., 2020).

Circular motion is a fundamental component of Kinematics, encompassing the movement of objects along a circular trajectory. Circular motion is commonly observed in many scenarios, from the celestial bodies revolving around the Sun to maneuvering a vehicle around a bend. An object undergoing circular motion maintains a consistent magnitude of speed while its direction undergoes continuous alteration, resulting in a variable velocity. The alteration in velocity gives rise to centripetal acceleration, which is oriented toward the center of the circular trajectory. Comprehending relative motion is crucial when multiple objects are in motion and their movements are observed from varying viewpoints. The concept of relative motion pertains to the movement exhibited by an entity when observed from a specific frame of reference. The relative motion between two independently moving objects is contingent upon the observer's motion. For instance, when considering the perspective of an observer, the relative velocities of two cars moving in different directions will vary. Non-inertial frames of reference are occasionally encountered, wherein accelerations give rise to apparent forces, also known as fictitious forces. Non-inertial frames refer to frames of reference that undergo acceleration, whether in the form of linear or angular motion. One illustration of a non-inertial frame can be observed when a car navigates a curved path. Within this particular context, the individuals aboard the vehicle experience a lateral force, commonly called the centrifugal force, which exerts an outward pull, directing them toward the outer edge of the curved trajectory (Kidder et al., 1996).

Nevertheless, this force is not genuine; rather, it arises as a result of the non-inertial characteristics of the reference frame. The field of robotics and automation heavily relies on the principles of Kinematics, as achieving accurate motion control is of utmost importance. The field of robotics encompasses the study of Kinematics, which pertains to analyzing and calculating the movement exhibited by robot arms and various other mechanical systems. The process of forward Kinematics involves determining the spatial coordinates and orientation of the end-effector, which refers to the specific component of the robot responsible for executing the assigned task. This determination is made by utilizing the joint angles as input parameters. In contrast, inverse Kinematics is a computational technique used to ascertain the specific joint angles necessary to attain the desired position and orientation of the end-effector. Using kinematic analyses enables engineers to devise robots capable of executing precise and efficient tasks (Lee, 1982).

Kinematics' breadth and practical uses are extensive and varied, rendering it an essential instrument in physics, engineering, and numerous scientific domains. Kinematics, through its systematic and quantitative methodology, assumes a crucial role in investigating and comprehending the natural realm and advancing sophisticated technologies that depend on accurate and regulated motion (Siciliano et al., 2009).

2.2. REFERENCE FRAMES AND DISPLACEMENT

To elucidate the movement of an entity, it is imperative to initially explicate its spatial location, denoting its specific whereabouts at any given moment. To provide a more precise description, it is necessary to specify the object's position about a suitable reference frame. The Earth is frequently employed as a frame of reference, and it is common practice to describe the position of objects about their distance from or proximity to the Earth. In mathematical terms, the representation of an object's position is commonly denoted by the variable x (Gómez et al., 2022).

2.2.1. Frames of Reference

To establish a position variable x, two choices must be made. The user is required to decide regarding the placement of $x = 0$ and the designation of the positive direction. Any frame is equally valid as long as consistency is maintained. However, it is not advisable to alter coordinate systems during a calculation.

Consider the scenario wherein an individual finds themselves seated within a train at a stationary station, only to observe an unexpected phenomenon wherein the station begins to exhibit a retrograde motion. Most individuals would assert that their failure to perceive the train's motion led them to perceive the station as being in motion. However, this observation demonstrates that selecting a coordinate system involves a third arbitrary decision, as valid frames of reference can exhibit variations by their relative motion.

Utilizing a coordinate system that moves about the Earth may initially appear unconventional. However, employing a frame of reference that moves in tandem with a train, for instance, can offer greater convenience when describing events occurring within the train. Considering frames of reference holds significant importance in describing an object's displacement (Bender & Beller, 2014).

Tip

When dealing with motion in classical mechanics, it is crucial to choose an appropriate reference frame from which we observe the motion. The choice of reference frame can significantly impact the way we describe and analyze the motion of objects. It's important to align the reference frame with the motion's characteristics to simplify calculations and accurately interpret the results.

2.2.2. Displacement

Displacement refers to the alteration in the spatial location of an entity about its designated frame of reference. As an illustration, when a car transitions from a residential dwelling to a retail establishment, its displacement can be defined as the relative spatial extent between the retail establishment and the reference frame, which in this case, is the residential dwelling. The term "displacement" conveys the idea that an object has undergone movement or has been shifted from its original position. Displacement refers to the alteration in the spatial location of an object and can be mathematically expressed as follows (Norgett et al., 1975):

$\Delta x = x_f - x_0$

where Δx represents displacement, fx represents the final position, and x_0 represents the initial position.

Figure 2.1 shows the significance of utilizing a frame of reference.

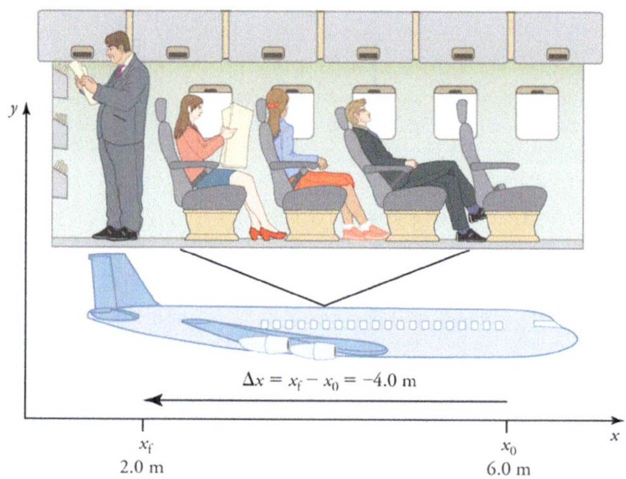

Figure 2.1. A passenger moves from his seat to the back of the plane (Source: Boundless, Creative Commons License)

2.3. SCALARS AND VECTORS

A vector is a mathematical entity with both magnitude and direction in mathematics and physics. On the other hand, a scalar is characterized solely by its magnitude, without any associated direction. The distinction between distance and displacement is rooted in their conceptualization and quantification within physics. Displacement is characterized by its direction and magnitude, whereas distance is determined solely by its magnitude. The notion of displacement functions as a visual representation of a vector quantity. Distance is a notable illustration of a scalar quantity. A vector is a class of physical quantity distinguished by its magnitude and direction. Additional illustrations of vectors encompass a velocity of 90 kilometers per hour in the eastward direction and a force of 500 newtons oriented vertically downward (Murphy, 1972).

Within mathematics, physics, and engineering, a vector is a geometric entity that exhibits attributes including magnitude, direction, and the ability to undergo vector addition by employing principles derived from vector algebra. Determining a vector's orientation in the context of motion in one dimension depends on the presence of a positive (+) or negative (−) sign. It is customary to depict a vector as a line segment with a clearly defined direction within academic contexts. The visual representation of the concept is frequently portrayed using an arrow that establishes a connection between an initial point A and a terminal point B, as exemplified in the provided illustration (Sekihara et al., 2004).

$\overrightarrow{AB}$

Within the realm of physics, a scalar is an essential physical entity that exhibits an inherent property of remaining unaltered when subjected to rotational motions or translations of the system of coordinates. A scalar is a mathematical entity in which a solitary numerical value can denote and exhibit magnitude without any directional aspect. Scalar quantities, such as temperature measurement at 20°C, do not possess a designated direction. Recognizing that a scalar can possess negative values is crucial, as demonstrated by a temperature measurement of -20°C. In this specific case, the negative symbol is used to indicate a location on a spectrum rather than indicating a particular direction. The utilization of arrows as a representation for scalars is nonexistent. Figure 2.2 depicts a visual representation that compares scalars and vectors (Pachava et al., 2020).

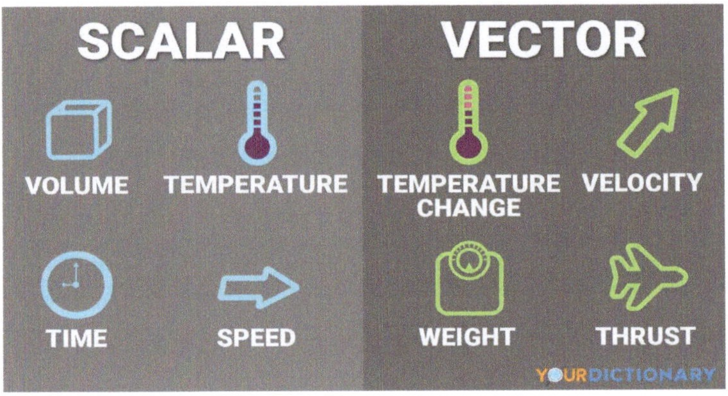

Figure 2.2. A brief list of quantities that are either scalars or vectors (Source: Your Dictionary, Creative Commons License)

2.4. SPEED AND VELOCITY

Physicists employ fundamental quantities of velocity and speed to describe the motion of objects about distance, time, and direction. As mentioned above, the two terms possess two separate and distinct connotations. However, it is common to observe these terms' interchangeable use. Therefore, what is the distinction? What is the reason for incorrectly using the words speed and velocity interchangeably (Yang et al., 2003)?

The underlying reasoning for this can be readily comprehended. Speed is the rate at which an object moves along a specific path over a certain period. On the other hand, velocity not only takes into account the speed of an object but also its direction of motion. In alternative terminology, speed is classified as a scalar magnitude, while velocity is defined as a vector magnitude. This concept can be exemplified by the differentiation between the expressions "50 km/hr" and "50 km/hr west." The former term refers to the rate of motion of a car along a road, whereas the latter term indicates the magnitude and direction of its motion (Xu et al., 2020).

The mathematical computation of speed is relatively uncomplicated, as it involves determining the average speed of an object by dividing the distance covered by the time taken to traverse the distance. The concept of velocity is mathematically intricate, and its calculation can vary depending on the available information regarding the object's motion. The average velocity calculation involves dividing the change in position (Δr) by the change in time (Δt) (Kafle et al., 2018).

2.4.1. Average Velocity

The categorization of average velocity as a vector quantity is a fundamental principle in the field of physics. The average velocity can be mathematically expressed as the ratio of the change in displacement (Δx) to the corresponding time intervals (Δt) during the displacement. The average velocity possesses either a positive or negative magnitude contingent upon the orientation of the displacement. The metric unit frequently utilized to quantify average velocity is meters per second (Figure 2.3) (SafarPour & Ghadiri, 2017).

To determine the average velocity, it is necessary to divide the total displacement by the elapsed time, as demonstrated by the subsequent equation:

$$\underline{v} = \frac{\Delta x}{\Delta t} = \frac{x_f - x_0}{t_f - t_0}$$

If the initial time is denoted as zero, then:

$$\underline{v} = \frac{\Delta x}{t}$$

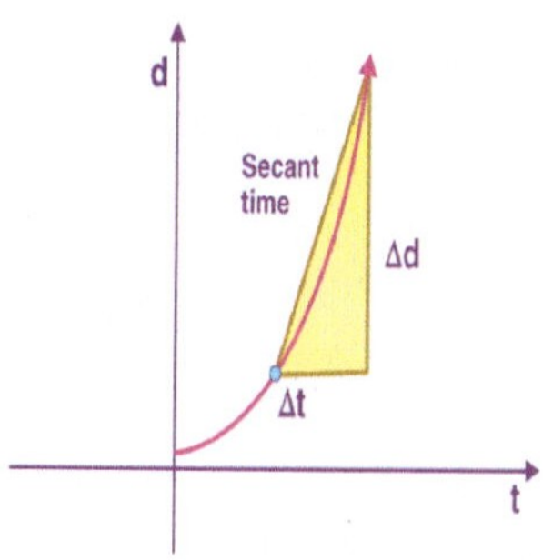

Figure 2.3. Schematic of the average velocity (Source: Byju's, Creative Commons License)

2.4.2. Instantaneous Velocity

In general, motion does not occur at a constant velocity or speed. When operating a motor vehicle, such as a car, it is common to experience a recurring pattern of acceleration and deceleration (Figure 2.4). A depiction of the motion regarding distance versus time would exhibit greater variability or curvature than a linear trend, suggesting motion characterized by a constant velocity (Noordhof et al., 2013).

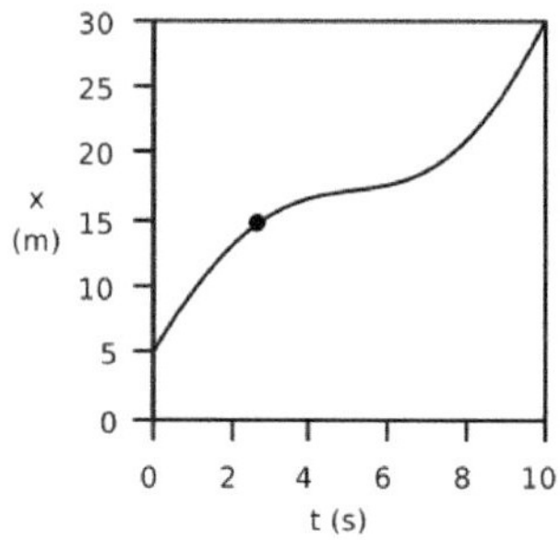

Figure 2.4. Motion with changing velocity (Source: LibreText, Creative Commons License)

Given the dynamic nature of velocity, it is possible to derive estimations for velocity through various approaches. One approach involves examining the instantaneous velocity, depicted as a single point on the graph illustrating the relationship between distance and time in a curved trajectory (Figure 2.5). To ascertain the velocity at a specific moment, it is necessary to calculate the slope at that particular point. It is necessary to identify a line that accurately represents the velocity at a given moment, as depicted graphically. The line in question corresponds to the tangent line of the curve at the given point. By extrapolating this line, one can readily compute the change in position concerning time and ascertain the velocity at the specified juncture. The instantaneous velocity of an object can be determined by calculating the slope of the tangent line that intersects the corresponding point on its position-time graph (Horan & Kavanagh, 2012).

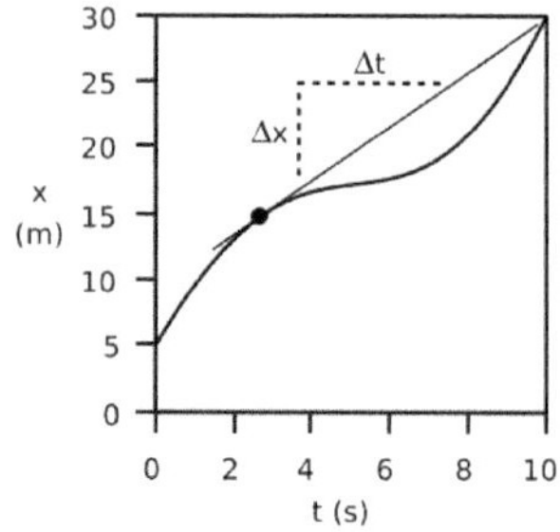

Figure 2.5. Determining instantaneous velocity (Source: LibreText, Creative Commons License)

2.5. ACCELERATION

Acceleration in physics refers to measuring the rate at which an object's velocity changes over a given period. A vector is a mathematical construct that possesses both magnitude and direction. Acceleration is intimately linked to the existence of a force, as expounded upon in Newton's Second Law. The force, being a vector quantity, is determined by multiplying the mass of an object undergoing acceleration and the acceleration vector, denoted as F = m. The meter per second squared (m/s^2) is the designated standard unit for measuring acceleration according to the International System of Units (SI) (Uzan, 2007). Acceleration is a vector quantity that exhibits parallelism with the direction of velocity change, although it may not necessarily align with the direction of motion. When an object undergoes a decrease in its velocity, commonly called deceleration, its acceleration is oriented in the opposite direction of its motion (Caldwell & Kamionkowski, 2009). The depiction of an object's motion in a visual form entails the construction of a graph that illustrates the object's position about time. The distance-time graph can be employed to generate an additional graph that depicts changes in velocity over time. The correlation between acceleration and velocity can be mathematically represented as the ratio of velocity, measured in meters per second, to time, measured in seconds. Therefore, it is feasible to infer a visual depiction of acceleration by analyzing a graph that illustrates the speed or position of an object (Liu & Fang, 2016).

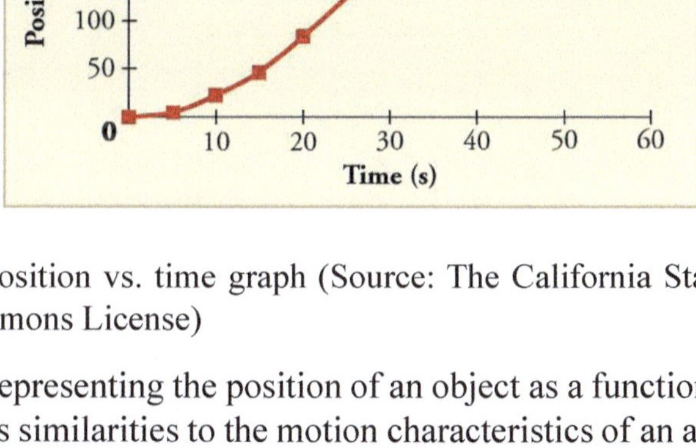

Figure 2.6. Position vs. time graph (Source: The California State University, Creative Commons License)

A graph representing the position of an object as a function of time. The graph exhibits similarities to the motion characteristics of an automobile. At the onset, the object's displacement gradually alters while it progressively acquires velocity. The velocity remains consistent at the midpoint while the displacement experiences a uniform alteration. As the object approaches its final stages, its velocity decreases, gradually altering its spatial location. Deducting a velocity versus time graph is possible based on the graphical representation provided (Figure 2.6) (Jones & Ellison, 1991).

It is observed that the object undergoes a gradual change in position during the initial phase of its journey, followed by an increasingly rapid change as it accelerates. The position subsequently decelerates towards the journey's conclusion, resulting in a slower rate of change. During the journey, when the velocity remains consistent, the position undergoes a uniform rate of change (Basson, 2002).

This graph illustrates the temporal evolution of the object's velocity. The object's velocity exhibits an initial increase due to acceleration at the onset, followed by a period of constancy in the middle, and ultimately decelerates towards the conclusion. It is important to observe that the graph provided provides a visual depiction of the derivative, specifically the slope, of the preceding position versus the time graph (Figure 2.7). The acceleration versus time graph can be derived from the provided graph (Taşar, 2010).

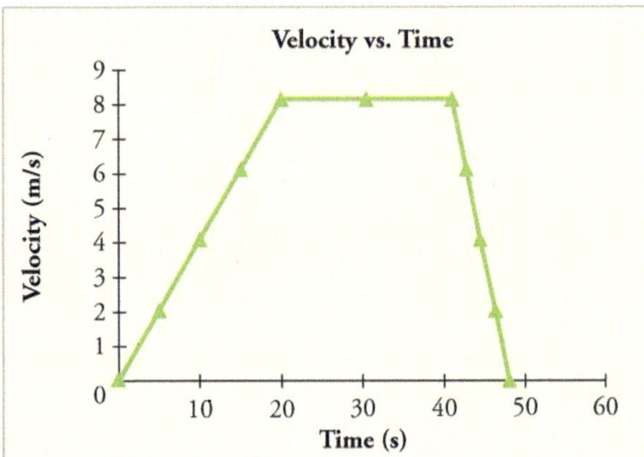

Figure 2.7. Velocity vs. time graph (Source: The California State University, Creative Commons License)

The object's velocity exhibits a positive correlation with its acceleration during the initial phase of the journey.

In addition, it is necessary to graph the slope of the velocity versus time plot. The graph depicts a consistent acceleration throughout the three distinct phases of motion. As previously mentioned, the object exhibits a gradual acceleration and displacement at the initial stages. According to the acceleration graph, it can be observed that the object experienced a consistent positive acceleration throughout the specified period (Guo et al., 2020). The acceleration was zero during the intermediate phase, wherein the object transitioned in position while maintaining a consistent velocity. This phenomenon occurs when the object reaches a state of equilibrium, where its velocity remains constant, and there is no further change in its motion. In the final stages of its motion, the object experiences a deceleration. The depicted value on the acceleration graph is presented as negative. It is important to observe that the object's motion remains in the forward direction (positive) in the illustration (Figure 2.8). However, due to its deceleration, the acceleration is characterized by a negative value (Gombosi et al., 2002).

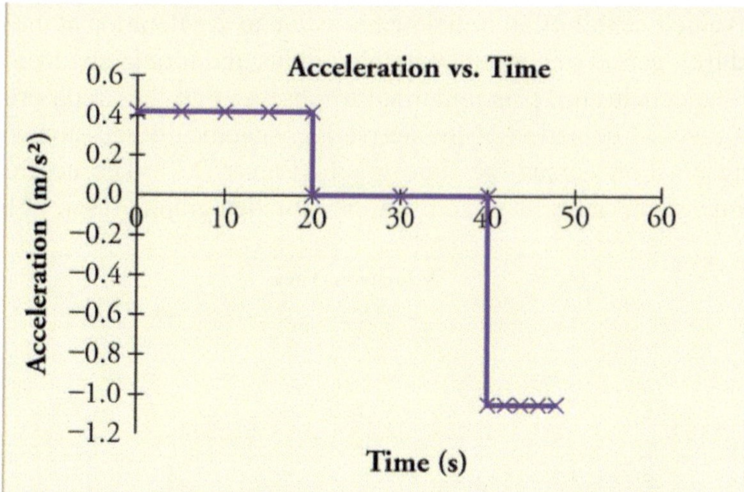

Figure 2.8. Acceleration vs. time graph (Source: The California State University, Creative Commons License)

The acceleration versus time graph indicates that the object experiences positive acceleration during the initial phase of its journey, increasing its speed. The object exhibits zero acceleration during its journey as it maintains a constant velocity. The object's acceleration is negative due to its deceleration towards the journey's conclusion (Figure 2.9).

Constant acceleration is observed when the velocity of an object undergoes uniform changes over equal intervals of time (Peng et al., 2018).

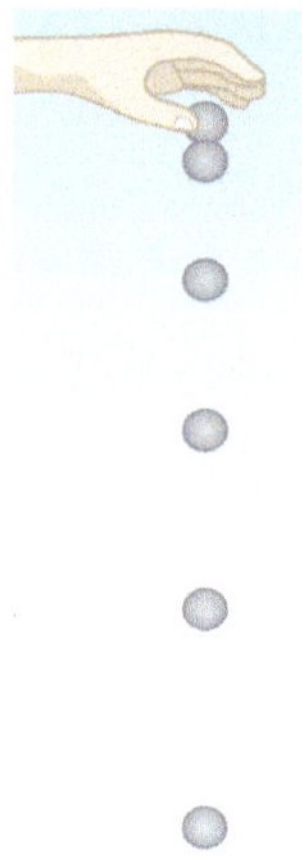

Figure 2.9. Constant acceleration occurs when an object's velocity changes by an equal amount in every equal time period (Source: Boundless, Creative Commons License)

In one-dimensional motion, releasing an object causes it to descend vertically toward the Earth's center, driven by the unchanging acceleration imparted by gravity.

An object undergoing uniform acceleration exhibits a consistent change in velocity, either increasing or decreasing, over a fixed duration of time. The determination of acceleration can be readily obtained through fundamental kinematic principles. The term "it" refers to the concept of acceleration, which is formally defined as the first derivative of velocity or, equivalently, the second derivative of location concerning time (Voth et al., 2002):

$a=\partial v/\partial t=\partial^2 x \partial t^2$

The assumption of constant acceleration does not significantly restrict the range of situations that can be studied, nor does it compromise the accuracy of the analysis. This is because a considerable number of scenarios exhibit constant acceleration. When constant acceleration is not applicable, one may analyze the system in distinct segments characterized by constant acceleration or calculate the average acceleration over a specified time interval (Derrick, 2004). The motion of objects in free fall can be described as a straightforward, unidimensional form of projectile motion characterized by the absence of horizontal displacement. As an illustration, when an individual extends their arm and releases a rock, the rock will solely descend in a vertical trajectory toward the Earth. If the rock were to be thrown rather than dropped, it would exhibit a trajectory resembling that of a projectile, akin to the trajectory observed when a ball is kicked. The kinematic equations establish a relationship between displacement, final velocity, initial velocity, acceleration, and time, owing to the algebraic properties of constant acceleration. Below is a summary of the equations (Brolinson et al., 2006).

$$X = x_0 + vt$$

$$V = v_0 + v/2$$

$$V = v_0 + at$$

$$X = x_0 + v_0 t + 1/2at^2$$

$$v^2 = v_0^2 + 2a(x - x_0)$$

2.6. PROJECTILE MOTION

The phenomenon of objects traversing through the atmosphere has captivated the attention of scientists and interested individuals across various epochs. Projectile motion is a fundamental concept in physics that reveals the intriguing characteristics of objects propelled into the air, influenced exclusively by the force of gravity.

The present article examines the phenomenon of projectile motion, elucidating its fundamental principles, delving into its mathematical expressions, and illuminating its practical implications in real-world contexts (Krist et al., 1993). A projectile refers to any object propelled into space solely under the influence of gravity. Gravity is the predominant force exerted on a projectile. This implies that while other forces may influence the object, their impact is relatively insignificant compared to the force of gravity. A projectile's trajectory refers to its path during its motion.

A baseball hit with a bat or thrown is a prime illustration of a projectile (Yabushita et al., 2007). When a particle is projected at an angle near the surface of the Earth, it follows a curved trajectory due to a constant acceleration directed towards the Earth's center. This assumption is made because the particle remains near the Earth's surface.

The trajectory followed by a particle of this nature is commonly referred to as a projectile, while the corresponding movement is known as projectile motion (Veeresha et al., 2021). The topic of discussion pertains to the acceleration experienced by a particle in both horizontal and vertical projectile motions. When a particle is launched into the atmosphere with a certain velocity, the sole force exerted on it throughout its duration in the atmosphere is the acceleration resulting from gravity (g). The acceleration in question is oriented in a vertical downward direction. In the horizontal direction, the absence of acceleration implies that the particle's velocity remains constant (Wee et al., 2012).

2.6.1. Parabolic Motion of Projectiles

Figure 2.10 shows the schematic diagram of the projectile motion.

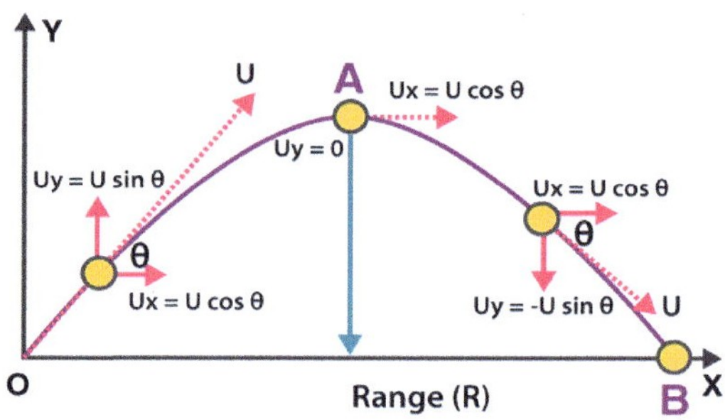

Figure 2.10. Illustration of projectile motion (Source: Byju's Creative Commons License)

Remember

In projectile motion, the horizontal and vertical motions of an object are independent of each other. The object's motion in the horizontal direction remains constant, while the vertical motion follows the effects of gravity. By treating these motions separately, we can break down complex problems into simpler components, making the analysis easier and more manageable.

2.6.2. Total Time of Flight

The vertical displacement(s) yields a resultant value of 0. Hence, the formula for calculating the time of flight is derived by utilizing the Equation of Motion:

gt2 = 2(uyt – sy)

Therefore, the time of flight is given by (Weibel et al., 2003):

$$ToalTimeofFlight(t) = \frac{2usinsin\Theta}{g}$$

2.6.3. Horizontal Range

Horizontal Range (OA) = Horizontal component of velocity (ux) × Total Flight Time (t)

R = u cos θ × 2u×sinθg

Therefore the Range is given by (Day, 2007):

$$HorizontalRange(R) = \frac{u^2 sinsin 2\Theta}{g}$$

2.6.4. Maximum Height of Projectile

Once the concept of a projectile has been understood, it would be advantageous to determine the maximum height achieved by the projectile above. The term "maximum height" pertains to the utmost vertical position attained by an object throughout its trajectory. The designation employed to denote the horizontal displacement of a projectile is commonly recognized as the range of the projectile. The range of a projectile is dependent on its initial velocity (Stewart, 2011).

Let v be the symbol used to represent the initial velocity, g be the variable denoting the gravitational acceleration, H stands for the maximum height, and θ is the angle of the initial velocity.The formula yields the maximum height attained by the projectile (Erlichson, 1983):

$$H = \frac{v_0^2 sin^2 q}{2g}$$

2.6.5. The Equation of Trajectory

$$EquationofTrajectory = xtantan\Theta - \frac{gx^2}{2u^2 cos^2\Theta}$$

The equation presented herein describes the trajectory in projectile motion, thereby substantiating the inherent parabolic nature of such motion (Shi et al., 2020).

2.6.6. Basketball Physics

Projectile motion is a form of motion that occurs in two dimensions, specifically within a plane. The prevailing assumption in projectile motion is that the sole force exerted on the projectile is attributable to the gravitational force. However, the definition of projectile motion in the real world remains a subject of inquiry. In what ways can the principles of projectile motion be observed and utilized in everyday situations? Allow us to examine tangible instances of projectile motion occurring within a two-dimensional framework (Changjan & Mueanploy, 2015).

It is widely acknowledged that basketball is a familiar subject to every one. The athlete executes a slight vertical displacement to successfully score a field goal, propelling the ball into the basket. The trajectory of the ball can be described as that of a projectile. Therefore, it is commonly known as projectile motion. What benefits does jumping confer upon an individual's likelihood of successfully scoring a basket? Upon their release, these objects experience the sole influence of the gravitational force. The force of gravity acts upon objects, causing them to experience a downward acceleration that is consistent and unbiased (Azhikannickal, 2019).

This suggests that the trajectory of an object thrown into the air can be accurately determined, allowing for the prediction of both its duration in flight and the distance it will travel from its initial launch point upon impact with the ground. In the absence of air resistance, the horizontal acceleration would be zero (Miller & Bartlett, 1996). This suggests that when a body is thrown near a surface, its motion can be regarded as two-dimensional, with acceleration occurring solely in one direction (Figure 2.11). However, it is imperative to ascertain how one can deduce that an object projected into the atmosphere adheres to a trajectory confined to two dimensions. To comprehend this concept, let us assume a spherical object that is in a state of motion, specifically rolling, as visually depicted in the illustration provided (Chen et al., 2009):

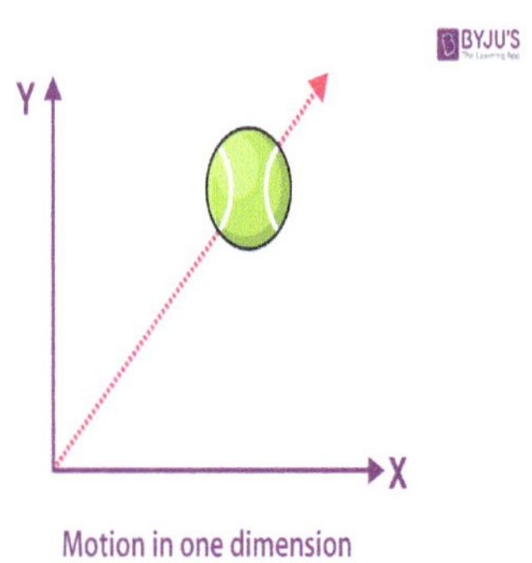

Figure 2.11. Motion in one dimension (Source: Byju's, Creative Commons License)

If the ball rolls along the depicted trajectory, what conclusions can be drawn regarding the dimensionality of its motion? The prevailing response would indicate that the object possesses both an x-component and a y-component, exhibiting motion within a two-dimensional plane. However, it is important to note that there is a discernible trajectory that can fully describe the motion of a basketball. Therefore, this serves as an illustration of unidirectional motion. Hence, the axis selection does not modify the inherent characteristics of the motion (Figure 2.12) (Tan & Miller, 1981).

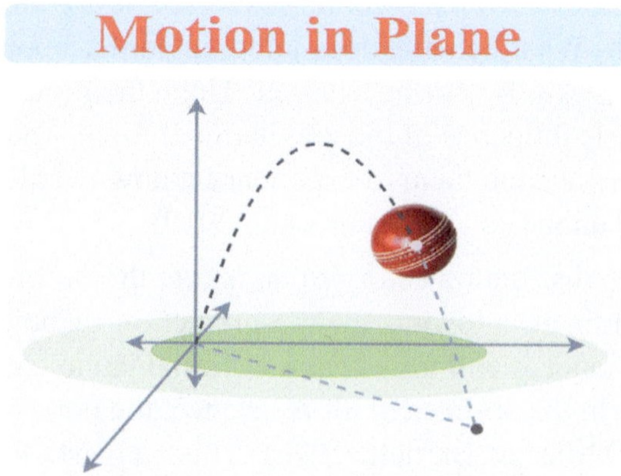

Figure 2.12. Motion in plane (Naimish Sahu, Creative Commons License)

When the ball is thrown at a given angle, as depicted, its velocity can be decomposed into three components: an x-component, a y-component, and a z-component. Therefore, does this imply that the motion being discussed is occurring in three dimensions? The observation made in this context is that a line is insufficient to describe or characterize a particular motion, whereas a plane can do so.

Hence, it can be concluded that a plane encompasses the entire trajectory for a projectile launched at any given angle. Therefore, it can be inferred that when an object is near the Earth's surface, and the impact of air resistance is negligible, its motion can be considered two-dimensional regardless of the chosen axes and the angle at which it is projected. If the axes are rotated in a manner such that they can fully determine the trajectory of the ball, as depicted in the illustration provided (Fontanella, 2006):

Therefore, it can be inferred that determining the dimension of a body's motion is contingent upon the smallest number of coordinates necessary to describe it fully (Okazaki & Rodacki, 2012).

2.7. UNIFORM CIRCULAR MOTION

The term "circular" denotes the movement that adheres to a curved path. Uniform circular motion is the motion of an object as it travels along a circular path. In this motion, the object covers the same distance along the circle's circumference within a constant time interval (Aly & Ahmed, 2020).

Circular motion is the term used to describe the path followed by an object as it travels along a circular trajectory. The concept of "Uniform Circular Motion" pertains to the motion of an object that maintains a constant velocity while following a circular path. In this particular situation, the magnitude of the speed remains constant while the direction of the velocity experiences changes (De la Fuente et al., 2017).

In the context of uniform circular motion, it has been empirically observed that the tangential speed exhibits a consistent and unvarying value at all locations along the circumference. Furthermore, it has been determined that the tangential velocity vector is oriented tangentially to the circumference at each point (Schott, 1937).

The tangential velocity is the vector tangent to the circular path and is directly related to the tangential speed. Consequently, the magnitude of the tangential velocity remains constant and is equal to the tangential speed in cases of uniform circular motion. The formula for determining the tangential is expressed as follows,

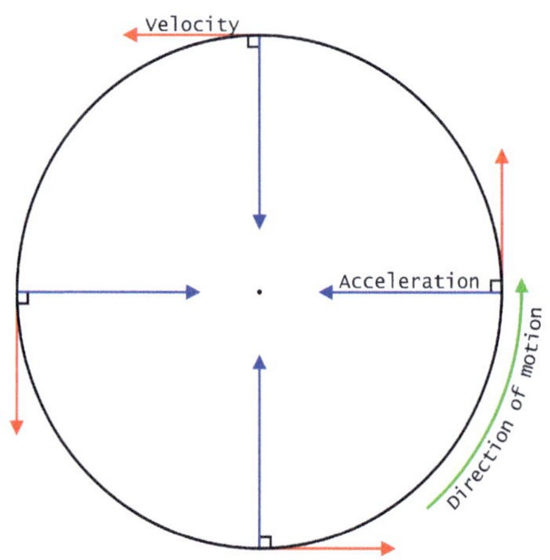

Figure 2.13. Illustration of uniform circular motion (Source: Toppr, Creative Commons License)

$v = 2\pi r / T$

The centripetal acceleration is,

$a = v^2 / r$

An angular velocity is possessed by any object engaged in a uniform circular motion (Figure 2.13) (Boyer, 1972).

When a particle is in circular motion, it is necessary for there to be a centripetal acceleration operating towards the center, which enables the particle to orbit around said center. The acceleration experienced by a particle is always perpendicular to its velocity, resulting in a change in direction but not magnitude. This acceleration characteristic leads to the particle's motion being classified as uniform circular motion. The term used to describe this phenomenon is centripetal acceleration, also referred to as radial acceleration. Additionally, the force exerted toward the center is called centripetal force (Westfall, 1972).

In the case of uniform circular motion, the acceleration is:

$a_r = v^2/r = \omega^2 r$

If the mass of the particle is m:

$F = ma$

$mv^2/r = m\omega^2 r$

The entity referred to as "special force" in the user's statement can be more accurately described as a force akin to tension or friction, potentially serving as a source for the generation of centripetal force. When vehicles navigate turns on roads, the frictional force between the tires and the ground generates the necessary centripetal force for turning (Klostergaard, 1976).

2.7.1. Centripetal Force

Centripetal force refers to the force applied to an object in motion along a curved trajectory, with its direction oriented towards the axis of rotation or the center of curvature.

The unit of measurement for the centripetal force is the Newton.

The centripetal force is consistently directed perpendicular to the object's displacement. Based on Newton's second law of motion, it can be inferred that the centripetal force acting on an object undergoing circular motion is consistently oriented toward the center of the circle (Johnson-Glenberg et al., 2016).

According to the centripetal force formula, the magnitude of centripetal force is determined by multiplying the mass of an object, measured in kilograms, by the square of its tangential velocity, measured in meters per second, and then dividing the result by the radius of the circular path, measured in meters. This suggests that doubling the tangential velocity will increase the resulting centripetal force by a factor of four. Mathematically speaking, it can be expressed as (Hubbard, 1995):

$$F = \frac{mv^2}{r}$$

Figure 2.14 shows illustration of centripetal force examples in daily life.

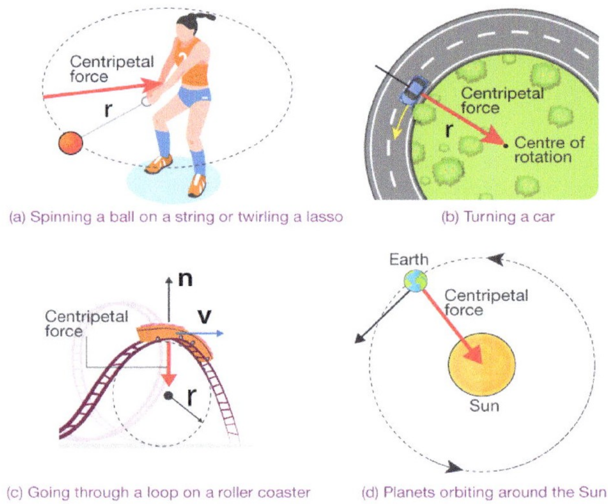

Figure 2.14. Illustration of different examples of centripetal force (Source: Byju's Creative Commons License)

1. When a ball is spun on a string or a lasso is twirled, the tension force exerted on the rope causes the object to be drawn toward the center (Pierratos et al., 2022).
2. The friction between the wheels and the ground generates the centripetal force required to turn a car.
3. When traversing a loop on a roller coaster, the centripetal force experienced by the rider is generated by the normal force exerted by the seat or wall, which propels the rider toward the center of the loop.

4. The planets in their orbits around the Sun experience a centripetal force that is attributed to the force of gravity (Vartholomeos & Papadopoulos, 2006).

2.7.2. Centrifugal Force

The centrifugal force is a theoretical force that emerges in a circular motion, manifesting itself along the radius and directed away from the circle's center. Force observation is not evident within an inertial frame of reference when conducting measurements (Schad et al., 2019). Newton is the designated unit of measurement for quantifying centrifugal force.

Centrifugal force is derived from the mathematical formulation of centripetal force, which effectively characterizes a physical phenomenon. Nevertheless, the centrifugal force is commonly regarded as fictitious due to its ability to alter the direction of the force (Spivack & Shure, 1985).

$$F = -\frac{mv^2}{r}$$

In this context, the symbol F represents the Centrifugal force, the symbol m denotes the object's mass, the symbol v signifies the speed or velocity of the object, and the symbol r represents the radius.

2.7.2.1. Centrifugal Force Examples in Daily Life

The phenomenon known as centrifugal force is observed to exert its influence on any object that is in motion along a circular trajectory, as perceived from a frame of reference that is itself rotating.

1. The weight experienced by an object at the poles and on the equator.
2. A bicycle executes a maneuver to change direction.
3. The operation of a vehicle while navigating a curved path (Hou et al., 2013).

2.8. RELATIVE MOTION

The initial introduction of reference frames was undertaken to address the topic of relative motion in either one or multiple dimensions. When referring to the velocity of an object, it is understood that this velocity is relative to a specific frame of reference. In daily activities, determining an object's velocity typically involves adopting the ground or Earth as the frame of reference. As an illustration, when one is aboard a moving train traveling at a velocity of 100 km/hr, the perceived velocity by another passenger situated within the same train would be zero (Somoza, 1998).

2.8.1. Relative Velocity

Let VA and VB represent the velocities of objects A and B about the ground, respectively. The relative velocity of A concerning B can be expressed as VAB = VA – VB. The equation for the relative velocity of object B concerning object A can be expressed as VBA = VB – VA

2.8.2. Relative Motion in One Dimension

In linear motion, objects exhibit movement along a single axis linearly. There exist only two conceivable scenarios:

1. The objects are exhibiting motion in a parallel direction.
2. The objects are exhibiting motion in a direction opposite to one another.

Consider the scenario of an individual seated within a train in motion, specifically traveling at a velocity of 100 kilometers per hour in the forward direction. According to the individual seated on the train, the trees outside the train's window are observed to recede at 100 kilometers per hour. This phenomenon occurs due to the observer's perspective, wherein the external surroundings appear to be moving in the opposite direction of the train at an equivalent velocity. In the context of various inquiries, when determining the velocity of entity A relative to entity B, it is customary to assume that entity B is stationary and subsequently assign the velocity of entity B to entity A in the opposing direction (Hibiki et al., 2004).

2.8.3. Relative Motion in Two Dimensions

The principle above can be extended to the realm of two-dimensional motion. To determine the velocity of object A relative to object B, it is assumed that object B is stationary. Consequently, the velocity of object B is assigned to object A in the opposite direction. Let us examine the motion of two objects, A and B, which possess velocities Va and Vb, respectively, about a shared frame of reference, such as the ground or the Earth (Curilef & Claro, 1997).

$$V_{ab} = v_a - v_b$$

Similarly, when considering object B's velocity relative to object A, it is assumed that object A is stationary, and the velocity of object A is assigned to object B in the opposite direction.

$$V_{ba} = v_b - v_a$$

In the context of a collision involving two bodies, it is always possible to describe the event within a frame of reference where the sum of the momenta of the bodies is zero. The frame above is commonly called the center-of-mass (or center-of-momentum) frame. In the collision above scenario involving two bodies of equal mass, it can be observed that the velocities of the two bodies are consistently equal in magnitude but opposite in direction. It is important to acknowledge that, within this particular frame of reference, the momenta of the outgoing particles exhibit antiparallel alignment rather than perpendicular alignment (Friemel et al., 1998).

Any assemblage of entities can be similarly characterized in a frame of reference where the total momentum is zero. The frame of reference is where the center of mass remains stationary. This fact can be readily observed by performing the process of differentiation on the following equation concerning time,

$$m\frac{dR}{dt} = \sum_{i=1}^{N} m_i \frac{dr_i}{dt}.$$

The summation of the momenta of all the bodies is represented on the right-hand side. The condition for the velocity of the center of mass, dR/dt, to be equal to zero results in an overall value of zero.

Suppose Newton's second law is valid in a particular frame of reference. In that case, it will also appear valid to an observer in motion with a constant velocity relative to that frame. The principle above, known as the principle of Galilean relativity, holds due to the perception of the moving observer, wherein a consistent velocity appears to be added to the velocity of each particle within the system. The alteration in velocity does not impact the particles' accelerations, as the additional velocity remains constant rather than experiencing acceleration. Consequently, the apparent force acting on each particle, determined by multiplying the mass by the acceleration, remains unaltered. Hence, it is acceptable to delineate a problem from the frame of reference of the center of momentum, given that the center of mass remains unaccelerated or from any other frame that moves at a constant velocity relative to it (Boffetta & Sokolov, 2002). If this principle is rigorously accurate, the fundamental forces of science ought not to possess any specific velocity. This assertion is deemed valid because the velocity of an object is subject to variation when observed from distinct yet equally valid frames of reference while the force exerted on the object remains constant. According to the theoretical framework proposed by James Clerk Maxwell, it is evident that the force laws governing electricity and magnetism possess an inherent speed, namely the speed of light (Sanliturk & Ewins, 1996).

2.9. MOTION IN NON-INERTIAL FRAMES

These studies acquired knowledge about acceleration, velocity, and displacement concepts. However, it is important to note that these quantities require a specific frame of reference to be accurately measured.

2.9.1. Types of Frame of Reference

Once a reference is chosen, it can be of two types:

1. Inertial frame of reference; and
2. Non-inertial frame of reference.

2.9.1.1. Inertial Frame of Reference

An inertial frame of reference can be defined as a coordinate system that satisfies the conditions necessary for Newton's laws of motion to hold. This statement suggests an object will maintain its rest or move at a constant velocity without external influences. Let us contemplate a hypothetical situation wherein an entity is positioned upon the Earth's terrestrial surface. From the vantage point of an observer situated on the surface of the Earth, the object exhibits the appearance of being motionless. Nevertheless, when observed from the Moon, the object exhibits apparent motion. In this context, it is crucial to ascertain the inertial frame of reference within this given scenario.

The notion of an inertial frame is intrinsically relative, as it necessitates the postulation of a reference frame to be designated as the inertial frame of reference. A more comprehensive and inclusive definition of an inertial frame can be articulated as follows: An inertial frame pertains to a frame of reference that remains stationary or moves with a consistent velocity about a presumed inertial reference frame (Delacroix et al., 2021).

2.9.1.2. Non-inertial Frame of Reference

A non-inertial frame can be characterized as one that experiences acceleration relative to the assumed inertial frame of reference. The applicability of Newton's law is invalid in these reference frames.

2.9.2. Pseudo Force

A pseudo force, also known as a fictitious force, inertial force, or d'Alembert force, is a force that appears to act on all masses when their motion is described within a non-inertial frame of reference, such as a rotating reference frame.

The concept of pseudo force arises when a reference frame experiences acceleration relative to a frame not undergoing acceleration.

The force denoted as F does not originate from any tangible interaction between two entities but from the acceleration denoted as 'a' of the non-inertial reference frame itself. Just as a frame can accelerate in any manner, pseudo forces also possess the capacity to be arbitrary, albeit solely in direct correlation to the frame's acceleration. Four pseudo forces are established in the context of frames experiencing an acceleration in commonly observed manners. These include a force resulting from the origin's rectilinear acceleration, two forces associated with rotation (namely, the Coriolis force and the Centrifugal force), and a fourth force known as the Euler force, which arises due to a varying rotational rate.For instance, when observing an individual positioned at a bus stop attentively observing a car in motion, it can be deduced that the individual is making an inference regarding the presence of a force acting upon the car, resulting in its acceleration. In this scenario, any issues are absent, and thus the utilization of the pseudo-force concept is unnecessary. However, when the individual in the car undergoing acceleration directs their gaze towards the individual positioned at the bus stop, they observe that the person appears to be experiencing acceleration about the car, despite the absence of any applied force. The concept of pseudo-force is necessary to transform a non-inertial frame of reference into an equivalent inertial frame of reference. Another example to consider involves a ball suspended from the roof of a train using a string incapable of being stretched. The string will assume a vertical orientation when the train is stationary or moving at a constant velocity along a straight path. A passenger can deduce that the net force exerted on the ball equals zero. If the train experiences acceleration, the string will exhibit an inclination relative to the vertical axis. The passenger experiences two non-collinear forces. However, the ball appears in equilibrium, assuming the train's acceleration remains constant. The inclusion of the concept of pseudo-force is necessary in this context (Claret & Venancio-Filho, 1991).

2.9.3. Coriolis Force

The Coriolis force is a perceived force that arises due to the rotation of the Earth. The deflection of winds towards the right in the northern hemisphere and toward the left in the southern hemisphere can be attributed to the Coriolis force. This phenomenon is commonly referred to as "Ferrel's Law." The magnitude of deflection increases as the wind velocity increases. The Coriolis force is exerted perpendicular to the pressure gradient force. The perpendicular orientation of the pressure gradient force concerning an isobar is observed. An increase in the pressure gradient force corresponds to a proportional increase in wind velocity and a greater magnitude of deflecting in the direction of the wind.

Did You Know?

The Coriolis effect is a fictitious force caused by the Earth's rotation, leading to deflection of moving objects such as air masses and ocean currents. It plays a significant role in large-scale weather patterns, like hurricanes and typhoons.

The Coriolis Effect is a perceptible phenomenon that arises from the rotation of a reference frame. The phenomenon arises when an object in motion along a linear trajectory is observed from a frame of reference that is not stationary. The Coriolis effect causes the apparent deviation of planes and long-distance air currents from a straight trajectory, resulting in a curved path around the Earth.The Coriolis Effect is a phenomenon that arises due to the rotation of the Earth. The Earth's rotational speed is greater at the equator than at the poles. Due to the Earth's wider circumference at the equator, the equatorial regions experience a rotational speed of approximately 1,600 kilometers per hour. The rotational speed is approximately 0.00008 kilometers per hour at the Earth's poles (Bouabid et al., 2013).

2.9.4. Fictitious Forces

The perceptible forces experienced within a moving vehicle, such as propelled into one's seat when the driver accelerates or laterally displaced during sharp turns, serve as commonplace illustrations of fictitious forces. Typically, these influences manifest without any discernible cause, except that the inherent frame of reference about a particular situation accelerates.

The concept of a "fictitious force" holds a specific definition within the framework of Newtonian mechanics, wherein it consistently exhibits a proportionality to the mass of the object upon which it exerts its influence. A notable illustration of such discernible influences can be observed in the hypothetical Coriolis force, which governs the deliberate precession, or circular rotation, of the plane of oscillation of a meticulously suspended pendulum. If a pendulum were to be suspended directly above the North Pole, it would exhibit a complete rotation of 360 degrees within 24 hours. When observed from a static vantage point in outer space, the pendulum would exhibit a consistent oscillation within a singular, unchanging plane as the Earth rotates beneath it. When observed from the perspective of outer space, it can be noted that there is an absence of any lateral force exerted on the pendulum, which would cause it to deviate from its swinging motion in a direction perpendicular to its plane of swing. Hence, the force in question is often labeled with the derogatory term "fictitious." Similarly, when inside a car, the sensation of being pushed back into the seat is not due to an actual force acting upon you (Engelhardt & Engelhardt, 2014). However, utilizing fictitious forces as a framework for analysis may be the most productive approach to comprehending the underlying dynamics of a given situation. Consider a cup of tea that has been stirred, which is an intriguing illustration of the effects of the Coriolis force. If a small number of tea leaves are introduced into a cup, they will accumulate in a concentrated manner at the central region of the cup's bottom surface. This behavior is contrary to the anticipated outcome of the hypothetical centrifugal force, which would suggest that the tea leaves would distribute themselves along the cup's periphery. When envisioning oneself rotating in synchronization with the agitated fluid, most of the fluid would seem stationary, whereas the cup would exhibit a counter-rotational motion relative to the observer. The rotational motion of the cup induces the entrainment of neighboring fluid. In the lower region of the cup, the fluid that is being dragged experiences the Coriolis force, which causes it to move toward the center of the cup. Consequently, the leaves in the fluid are also carried in the same direction. Albert Einstein successfully eliminated the enduring dichotomy between genuine and imaginary forces through his theory of general relativity. General relativity is Albert Einstein's theory of gravity, widely regarded as the quintessential example of a fundamental force. However, the fundamental principle of Einstein's theory posits that gravity is a fictitious force, or more precisely, indistinguishable from a fictitious force (Kumar et al., 2018).

Activity 1.1

Imagine throwing a ball from the edge of a tall building. Discuss how the ball's horizontal and vertical motions change as it travels through the air. What factors influence the time of flight and the distance covered by the ball horizontally? Calculate the ball's maximum height and the time taken to reach it.

SUMMARY

This chapter provides a comprehensive introduction to kinematics, the study of motion without considering its underlying causes. It starts by discussing reference frames and displacement, emphasizing the significance of choosing appropriate frames for accurate analysis. The distinction between scalars and vectors is highlighted, with examples of quantities falling under each category. Speed and velocity, key components of motion, are explored, along with the differences between average and instantaneous velocity. The concept of acceleration as the rate of change of velocity is introduced, demonstrating its relevance in describing motion. Projectile motion is thoroughly examined, particularly the parabolic trajectory of projectiles and various related parameters like total time of flight, horizontal range, and maximum height. Real-world applications, such as basketball physics, demonstrate the practical importance of understanding projectile motion. Uniform circular motion is explored, emphasizing the centripetal and centrifugal forces acting on an object in circular motion. Relative motion in different frames of reference is discussed, and the topic of motion in non-inertial frames is introduced, encompassing pseudo forces like the Coriolis force. This chapter provides a solid foundation for understanding the fundamental principles of kinematics, setting the stage for further exploration into the dynamics of motion and its various applications in classical mechanics.

Review Questions

1. What is the significance of choosing an appropriate reference frame in analyzing motion? Provide an example to illustrate its importance.
2. Differentiate between scalar and vector quantities in kinematics. Give two examples of each type.

3. Define average velocity and instantaneous velocity. How are they different from each other? Provide an example to illustrate the distinction.
4. Explain the concept of centripetal force in uniform circular motion. How does it influence an object's path?
5. Describe the motion of a projectile launched at an angle to the ground. What are the key parameters affecting its trajectory?
6. Discuss the independence of horizontal and vertical motions in projectile motion. How does this property simplify the analysis of projectile trajectories?
7. What is relative motion? Provide an example to demonstrate how observers in different reference frames describe the same motion differently.
8. Explain the notion of pseudo force in non-inertial frames of reference. When and why do we consider pseudo forces in such frames?

REFERENCES

1. Aly, A. M., & Ahmed, S. E. (2020). Effects of uniform circular motion on natural convection in a cavity filled with a nanofluid using incompressible SPH method. *International Communications in Heat and Mass Transfer*, *116*, 104646.
2. Azhikannickal, E. (2019). Sports, smartphones, and simulation as an engaging method to teach projectile motion incorporating air resistance. *The Physics Teacher*, *57*(5), 308–311.
3. Basson, I. (2002). Physics and mathematics as interrelated fields of thought development using acceleration as an example. *International Journal of Mathematical Education in Science and Technology*, *33*(5), 679–690.
4. Bender, A., & Beller, S. (2014). Mapping spatial frames of reference onto time: A review of theoretical accounts and empirical findings. *Cognition*, *132*(3), 342–382.
5. Boffetta, G., & Sokolov, I. M. (2002). Statistics of two-particle dispersion in two-dimensional turbulence. *Physics of Fluids*, *14*(9), 3224–3232.
6. Bouabid, M. P., Dupret, M. A., Salmon, S., Montalbán, J., Miglio, A., & Noels, A. (2013). Effects of the Coriolis force on high-order g modes in γ Doradus stars. *Monthly Notices of the Royal Astronomical Society*, *429*(3), 2500–2514.
7. Boyer, T. H. (1972). Mass renormalization and radiation damping for a charged particle in uniform circular motion. *American Journal of Physics*, *40*(12), 1843–1846.
8. Brolinson, P. G., Manoogian, S., McNeely, D., Goforth, M., Greenwald, R., & Duma, S. (2006). Analysis of linear head accelerations from collegiate football impacts. *Current Sports Medicine Reports*, *5*, 23–28.
9. Caldwell, R. R., & Kamionkowski, M. (2009). The physics of cosmic acceleration. *Annual Review of Nuclear and Particle Science*, *59*, 397–429.
10. Changjan, A., & Mueanploy, W. (2015, May). Projectile motion in real-life situation: Kinematics of basketball shooting. In *Journal of Physics: Conference Series*. IOP Publishing. (Vol. 3, pp. 1200–1206).
11. Chen, H. T., Tien, M. C., Chen, Y. W., Tsai, W. J., & Lee, S. Y. (2009). Physics-based ball tracking and 3D trajectory reconstruction with applications to shooting location estimation in basketball video.

Journal of Visual Communication and Image Representation, *20*(3), 204–216.

12. Chen, Z., Wang, Z., Liang, R., Liang, B., & Zhang, T. (2020). Virtual-joint based motion similarity criteria for human–robot kinematics mapping. *Robotics and Autonomous Systems*, *125*, 103412.
13. Claret, A. M., & Venancio-Filho, F. (1991). A modal superposition pseudo-force method for dynamic analysis of structural systems with non-proportional damping. *Earthquake Engineering & Structural Dynamics*, *20*(4), 303–315.
14. Curilef, S., & Claro, F. (1997). Dynamics of two interacting particles in a magnetic field in two dimensions. *American Journal of Physics*, *65*(3), 244–250.
15. Day, L. H. (2007). Can physics help win the case? A real-world application of an unusual two-dimensional projectile motion situation. *Physics Education*, *42*(6), 621.
16. De la Fuente, D., Romero, A., & Torres, P. J. (2017). Uniform circular motion in general relativity: Existence and extendibility of the trajectories. *Classical and Quantum Gravity*, *34*(12), 125016.
17. Delacroix, B., Rastoueix, J., Fradette, L., Bertrand, F., & Blais, B. (2021). CFD-DEM simulations of solid-liquid flow in stirred tanks using a non-inertial frame of reference. *Chemical Engineering Science*, *230*, 116137.
18. Derrick, T. R. (2004). The effects of knee contact angle on impact forces and accelerations. *Medicine and Science in Sports and Exercise*, *36*(5), 832–837.
19. Engelhardt, N. T., & Engelhardt, N. (2014). Gravity in Gravity's Rainbow–Force, Fictitious Force, and Frame of Reference; or: The Science and Poetry of Sloth. *Orbit: A Journal of American Literature*, *2*(2), 123–125.
20. Erlichson, H. (1983). Maximum projectile range with drag and lift, with particular application to golf. *American Journal of Physics*, *51*(4), 357–362.
21. Fontanella, J. J. (2006, December). Some Aspects of the Physics of Shooting a Basketball. In *American Astronomical Society Meeting Abstracts* (Vol. 209, pp. 68–01).
22. Friemel, B. H., Bohs, L. N., Nightingale, K. R., & Trahey, G. E. (1998). Speckle decorrelation due to two-dimensional flow gradients. *IEEE*

Transactions on Ultrasonics, Ferroelectrics, and Frequency Control, 45(2), 317–327.

23. Gombosi, T. I., Tóth, G., De Zeeuw, D. L., Hansen, K. C., Kabin, K., & Powell, K. G. (2002). Semirelativistic magnetohydrodynamics and physics-based convergence acceleration. *Journal of Computational Physics, 177*(1), 176–205.
24. Gómez, D. D., Bevis, M. G., & Caccamise, D. J. (2022). Maximizing the consistency between regional and global reference frames utilizing inheritance of seasonal displacement parameters. *Journal of Geodesy, 96*(2), 9.
25. Gould, K., Lovejoy, A. E., Jegley, D., Neal, A. L., Linton, K. A., Bergan, A. C., & Bakuckas Jr, J. G. (2015). Nonlinear Analysis and Experimental Behavior of a Curved Unitized Stitched Panel. *Journal of Aircraft, 52*(2), 628–637.
26. Guo, F., Liu, Y. H., Li, X., Li, H., Daughton, W., & Kilian, P. (2020). Recent progress on particle acceleration and reconnection physics during magnetic reconnection in the magnetically-dominated relativistic regime. *Physics of Plasmas, 27*(8), 6541–6542.
27. Hibiki, T., Takamasa, T., & Ishii, M. (2004, January). One-dimensional drift-flux model and constitutive equations for relative motion between phases in various two-phase flow regimes at microgravity conditions. In *International Conference on Nuclear Engineering* (Vol. 4689, pp. 377–386).
28. Horan, S. A., & Kavanagh, J. J. (2012). The control of upper body segment speed and velocity during the golf swing. *Sports Biomechanics, 11*(2), 165–174.
29. Hou, H. W., Warkiani, M. E., Khoo, B. L., Li, Z. R., Soo, R. A., Tan, D. S. W., ... & Lim, C. T. (2013). Isolation and retrieval of circulating tumor cells using centrifugal forces. *Scientific Reports, 3*(1), 1259.
30. Hubbard, T. L. (1995). Environmental invariants in the representation of motion: Implied dynamics and representational momentum, gravity, friction, and centripetal force. *Psychonomic Bulletin & Review, 2*, 322–338.
31. Johnson-Glenberg, M. C., Megowan-Romanowicz, C., Birchfield, D. A., & Savio-Ramos, C. (2016). Effects of embodied learning and digital platform on the retention of physics content: Centripetal force. *Frontiers in Psychology, 7*, 1819.

32. Jones, F. C., & Ellison, D. C. (1991). The plasma physics of shock acceleration. *Space Science Reviews*, *58*(1), 259–346.
33. Kafle, P. R., Sharma, S., Lewis, G. F., Robotham, A. S., & Driver, S. P. (2018). The need for speed: escape velocity and dynamical mass measurements of the Andromeda galaxy. *Monthly Notices of the Royal Astronomical Society*, *475*(3), 4043–4054.
34. Kidder, S. M., Abuzzahab, F. S., Harris, G. F., & Johnson, J. E. (1996). A system for the analysis of foot and ankle kinematics during gait. *IEEE transactions on rehabilitation engineering*, *4*(1), 25–32.
35. Klostergaard, H. (1976). Determination of gravitational acceleration g using a uniform circular motion. *American Journal of Physics*, *44*(1), 68–69.
36. Krist, H., Fieberg, E. L., & Wilkening, F. (1993). Intuitive physics in action and judgment: The development of knowledge about projectile motion. *Journal of Experimental Psychology: Learning, Memory, and Cognition*, *19*(4), 952.
37. Kumar, B. S., Kalpathy, S. K., & Anandhan, S. (2018). Synergism of fictitious forces on nickel cobaltite nanofibers: electrospinning forces revisited. *Physical Chemistry Chemical Physics*, *20*(7), 5295–5304.
38. Lee, C. G. (1982). Robot arm kinematics, dynamics, and control. *Computer*, *15*(12), 62–80.
39. Liu, G., & Fang, N. (2016). Student misconceptions about force and acceleration in physics and engineering mechanics education. *International Journal of Engineering Education*, *32*(1), 19–29.
40. Miller, S., & Bartlett, R. (1996). The relationship between basketball shooting kinematics, distance and playing position. *Journal of sports sciences*, *14*(3), 243–253.
41. Murphy, A. H. (1972). Scalar and vector partitions of the probability score: Part I. Two-state situation. *Journal of Applied Meteorology (1962–1982)*, 273–282.
42. Noordhof, D. A., Foster, C., Hoozemans, M. J., & De Koning, J. J. (2013). Changes in speed skating velocity in relation to push-off effectiveness. *International Journal of Sports Physiology and Performance*, *8*(2), 188–194.
43. Norgett, M. J., Robinson, M. T., & Torrens, I. M. (1975). A proposed method of calculating displacement dose rates. *Nuclear Engineering and Design*, *33*(1), 50–54.

44. Okazaki, V. H. A., & Rodacki, A. L. F. (2012). Increased distance of shooting on basketball jump shot. *Journal of Sports Science & Medicine*, *11*(2), 231.
45. Pachava, S., Dharmavarapu, R., Vijayakumar, A., Jayakumar, S., Manthalkar, A., Dixit, A., ... & Bhattacharya, S. (2020). Generation and decomposition of scalar and vector modes carrying orbital angular momentum: a review. *Optical Engineering*, *59*(4), 041205–041205.
46. Peng, Y., Deng, B., Zhang, J., Geng, F., Qin, W., & Liu, L. (2018). Anderson acceleration for geometry optimization and physics simulation. *ACM Transactions on Graphics (TOG)*, *37*(4), 1–14.
47. Pierratos, T., Sotirios, M., & Eirini, D. (2022). Experimental investigation of the factors that affect the magnitude of the centripetal force exerted on a rotating body in a flipped classroom. *Physics Education*, *57*(3), 035012.
48. SafarPour, H., & Ghadiri, M. (2017). Critical rotational speed, critical velocity of fluid flow and free vibration analysis of a spinning SWCNT conveying viscous fluid. *Microfluidics and Nanofluidics*, *21*, 1–23.
49. Saltiel, E., & Malgrange, J. L. (1980). 'Spontaneous' ways of reasoning in elementary kinematics. *European Journal of Physics*, *1*(2), 73.
50. Sanliturk, K. Y., & Ewins, D. J. (1996). Modelling two-dimensional friction contact and its application using harmonic balance method. *Journal of Sound and Vibration*, *193*(2), 511–523.
51. Schad, J., Lewis, M. W., & Smith, W. K. (2019). Quo vadis, paradox? Centripetal and centrifugal forces in theory development. *Strategic Organization*, *17*(1), 107–119.
52. Schott, G. A. (1937). The uniform circular motion with invariable normal spin of a rigidly and uniformly electrified sphere, IV. *Proceedings of the Royal Society of London. Series A-Mathematical and Physical Sciences*, *159*(899), 570–591.
53. Sekihara, K., Nagarajan, S. S., Poeppel, D., & Marantz, A. (2004). Asymptotic SNR of scalar and vector minimum-variance beamformers for neuromagnetic source reconstruction. *IEEE Transactions on Biomedical Engineering*, *51*(10), 1726–1734.
54. Shi, Y., Hua, Y., & Pan, G. (2020). Experimental study on the trajectory of projectile water entry with asymmetric nose shape. *Physics of Fluids*, *32*(12).

55. Siciliano, B., Sciavicco, L., Villani, L., & Oriolo, G. (2009). Kinematics. *Robotics: Modelling, Planning and Control*, 5(5), 39–103.
56. Somoza, R. (1998). Updated azca (Farallon)—South America relative motions during the last 40 My: implications for mountain building in the central Andean region. *Journal of South American Earth Sciences*, *11*(3), 211–215.
57. Spivack, G., & Shure, M. B. (1985). ICPS and beyond: Centripetal and centrifugal forces. *American Journal of Community Psychology*, *13*(3), 226.
58. Stewart, S. M. (2011). Comment on 'On the locus formed by the maximum heights of projectile motion with air resistance'. *European Journal of Physics*, *32*(2), L7.
59. Tan, A., & Miller, G. (1981). Kinematics of the free throw in basketball. *American Journal of Physics*, *49*(6), 542–544.
60. Taşar, M. F. (2010). What part of the concept of acceleration is difficult to understand: the mathematics, the physics, or both?. *ZDM*, *42*, 469–482.
61. Uzan, J. P. (2007). The acceleration of the universe and the physics behind it. *General Relativity and Gravitation*, *39*, 307–342.
62. Vartholomeos, P., & Papadopoulos, E. (2006, May). Analysis, design and control of a planar micro-robot driven by two centripetal-force actuators. In *Proceedings 2006 IEEE International Conference on Robotics and Automation, 2006. ICRA 2006.* IEEE. (Vol. 2, pp. 649–654).
63. Veeresha, P., Ilhan, E., & Baskonus, H. M. (2021). Fractional approach for analysis of the model describing wind-influenced projectile motion. *Physica Scripta*, *96*(7), 075209.
64. Voth, G. A., La Porta, A., Crawford, A. M., Alexander, J., & Bodenschatz, E. (2002). Measurement of particle accelerations in fully developed turbulence. *Journal of Fluid Mechanics*, *469*, 121–160.
65. Waldron, K. J., & Schmiedeler, J. (2016). Kinematics. *Springer Handbook of Robotics*, 4, 11–36.
66. Wee, L. K., Chew, C., Goh, G. H., Tan, S., & Lee, T. L. (2012). Using Tracker as a pedagogical tool for understanding projectile motion. *Physics Education*, *47*(4), 448.

67. Weibel, D., Wong, S., Lockyer, N., Blenkinsopp, P., Hill, R., & Vickerman, J. C. (2003). A C60 primary ion beam system for time of flight secondary ion mass spectrometry: its development and secondary ion yield characteristics. *Analytical Chemistry*, *75*(7), 1754–1764.
68. Westfall, R. S. (1972). Circular motion in seventeenth-century mechanics. *Isis*, *63*(2), 184–189.
69. Xu, T., Zhang, S., Jiang, Z., Liu, Z., & Cheng, H. (2020). Collision avoidance of high-speed obstacles for mobile robots via maximum-speed aware velocity obstacle method. *IEEE Access*, *8*, 138493–138507.
70. Yabushita, K., Yamashita, M., & Tsuboi, K. (2007). An analytic solution of projectile motion with the quadratic resistance law using the homotopy analysis method. *Journal of Physics A: Mathematical and Theoretical*, *40*(29), 8403.
71. Yang, V. X., Gordon, M. L., Qi, B., Pekar, J., Lo, S., Seng-Yue, E., ... & Vitkin, I. A. (2003). High speed, wide velocity dynamic range Doppler optical coherence tomography (Part I): System design, signal processing, and performance. *Optics Express*, *11*(7), 794–809.
72. Zhang, A., Wei, J., Shi, L., Qin, D., & Lim, T. C. (2019). Modeling and dynamic response of parallel shaft gear transmission in non-inertial system. *Nonlinear Dynamics*, *98*, 997–1017.

CHAPTER 3

Dynamics: Forces and Motion

Contents

3.1. INTRODUCTION

The phenomenon of motion captures our attention. The aesthetic appeal of motion is evident in various captivating instances, such as the graceful leap of a dolphin, the agile movements of a pole vaulter, the elegant flight of a bird, or the precise trajectory of a satellite in orbit. These remarkable motion displays inspire awe as we contemplate the underlying forces required to achieve such spectacles. As a field of study, kinematics pertains to the analysis of motion. However, it is important to note that kinematics solely focuses on elucidating the characteristics of an object's movement, specifically its velocity and acceleration. The field of dynamics encompasses examining the forces that influence the motion of objects and systems in motion. The fundamental principles governing the motion of objects, known as Newton's laws of motion, serve as the cornerstone of the field of dynamics. These laws exemplify the wide range and straightforwardness of the principles that govern the functioning of nature. Moreover, these laws are universal, as they apply to analogous circumstances on Earth and in outer space (Nestler & Voigt, 2021).

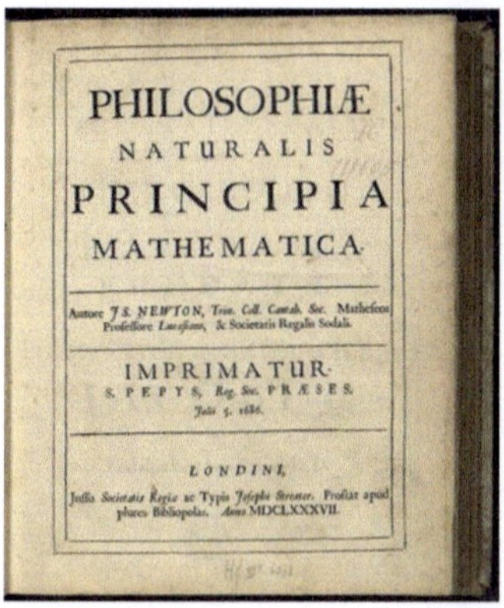

Figure 3.1. Issac Newton's monumental work, Philosophiae Naturalis Principia Mathematica, was published in 1687 (Source: Joint documentation service of the University of Strasbourg, Creative Commons License)

The laws of motion formulated by Isaac Newton (1642–1727) constitute a significant component of his extensive work, contributing to his esteemed reputation. The formulation of Newton's laws represents a pivotal shift from the Renaissance to the modern era. This transition period was marked by a paradigm shift in individuals' perceptions of the physical cosmos. Throughout numerous centuries, scholars in the field of natural philosophy engaged in extensive discussions regarding the fundamental characteristics of the universe. These deliberations heavily relied on logical principles, with significant emphasis on the ideas of esteemed classical philosophers, notably Aristotle (384–322 BC). The significant contributions of renowned intellectuals such as Newton and Galileo facilitated the change in question (Figure 3.1) (Moo et al., 2018).

Galileo played a pivotal role in establishing observation as the definitive criterion for determining truth, surpassing the reliance on "logical" argumentation. One of Galileo's most significant accomplishments was his utilization of the telescope, which effectively showcased the criticality of observation. He made significant astronomical discoveries, such as identifying the presence of moons in orbit around Jupiter and conducted observations that contradicted certain ancient concepts and religious doctrines. Due to this rationale and Galileo's approach to handling individuals in positions of power, he underwent a trial by the Inquisition and faced subsequent consequences. During the concluding years of his existence, he was subjected to a state of confinement within his residence. Due to the existence of prior individuals who had also made significant observations regarding the nature of the universe and the subsequent confirmation of Galileo's findings through repeated observations, it became untenable to suppress or refute his work. Following his demise, the work produced by the individual in question underwent rigorous scrutiny from various scholars, resulting in the eventual recognition and acceptance of his ideas by both ecclesiastical and scientific establishments (Sjökvist et al., 2017).

3.2. DEVELOPMENT OF FORCE CONCEPT

The study of the interactions between the forces responsible for the motion of different objects and systems is the focus of the academic field known as dynamics. Establishing a functional definition of force is necessary to have any hope of understanding this concept.

The first place to look for a point of reference is our intuitive understanding of force, which incorporates applying either a pulling or pushing force. It is good knowledge that a force has both magnitude and direction, which makes it a vector quantity. Furthermore, it is possible for a force's magnitude and direction to display a large amount of variation. For example, during a cannonball's ascent into the atmosphere, a cannon applies a substantial amount of force to the projectile as it travels through it (Hestenes & Halloun, 1995).

On the other hand, the gravitational pull that the Earth has on a flea is not even close to being noticeable. Because of the variety of experiences we have daily, we have a deep comprehension of the combined effects of several different forces. It is reasonable to assume that the resultant force will be aligned with the direction indicated in Figure 3.2 when two individuals apply forces on a third individual moving in opposite directions, as shown in the diagram. Given that force is a vector quantity, it adheres to the rules of vector addition, which are illustrated in Figure 3.2(a) in the case of two ice skaters moving in opposite directions. In the same way that arrows represent other vectors, forces can be coupled using either the well-known head-to-tail method or approaches based on trigonometry. The Two-Dimensional Kinematics framework served as the basis for developing the ideas being discussed in this discourse (Coletta et al., 2007).

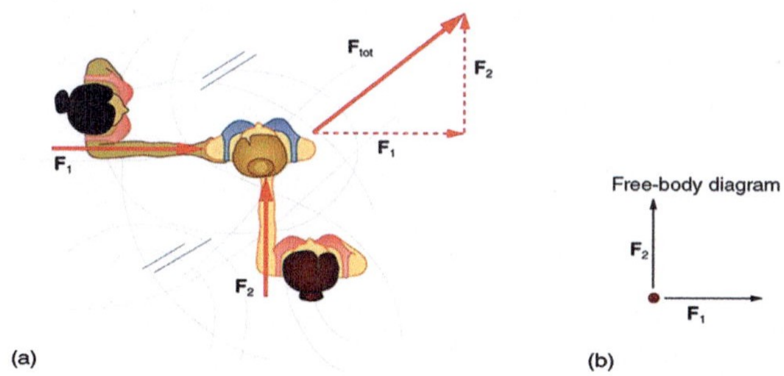

Figure 3.2. Part (a) shows an overhead view of two ice skaters pushing on a third. In part (b), we see a free-body diagram representing the forces acting on the third skater (Source: Rice University, Creative Commons License).

The first example of a free-body diagram is depicted in Figure 3.2(b), showcasing a technique employed to represent the external forces exerted on a physical entity visually. A solitary, unconnected point, a free body, denotes the physical entity under consideration. Solely the forces exerted on the body from external sources, known as external forces, are depicted. The forces depicted are exclusively those that have been illustrated, as solely external forces exerted on the object impact its motion. Internal forces within the body can be disregarded (Epstein, 2007).

A more rigorous conceptualization of force can be established by establishing a quantitative framework that relies on a standardized force. This is similar to how distance is measured using units relative to a standardized distance. One potential approach involves the extension of a spring by a predetermined and constant distance, as depicted in Figure 3.3. The subsequent utilization of the spring's restorative force, which enables it to return to its original shape, serves as a benchmark. The magnitudes of all other forces can be expressed as multiples of this standardized unit of force. There are numerous alternative options available for standard forces (Planinic et al., 2010).

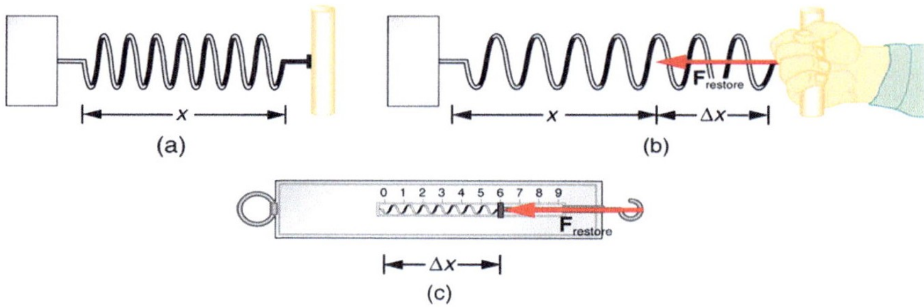

Figure 3.3. The force exerted by a stretched spring can be used as a standard unit of force. (a) This spring has a length x when undistorted. (b) When stretched a distance (Source: Openstax, Creative Commons License).

The displacement, denoted as Δx, results in the spring exerting a restoring force, Frestore, that can be replicated. The utilization of a spring to quantify force is a characteristic feature of a spring scale. The restorative force, denoted as F_restore, is applied to any object affixed to the hook. The magnitude of the force standard employed is 6 units for the restoration force (Bruun & Brewe, 2013).

3.3. NEWTON'S FIRST LAW OF MOTION: INERTIA

Empirical evidence suggests that an object currently in a state of rest will remain in that state unless it is subjected to external pressures. On the other hand, empirical evidence indicates that an object currently in motion will slow down and eventually stop if external influences do not counteract it. The first law of motion that Newton proposed, on the other hand, can be stated as follows:

The first law of motion that Newton proposed states that an object initially at rest will continue to remain at rest. In contrast, an object already in motion will continue to move at a constant velocity until it comes under the influence of an external force (Saglam-Arslan & Devecioglu, 2010).

It is clear that the verb "remains" has been used multiple times throughout the sentence. One way to think about this law is as a mechanism for preserving the existing motion.

Remember

An object at rest stays at rest and an object in motion stays in motion with the same speed and in the same direction unless acted upon by an unbalanced force – this is the principle of inertia.

According to the first law of motion posited by Isaac Newton, any change in velocity, regardless of whether it is a change in magnitude or direction, requires the presence of a cause, more precisely, a net external force. This is a prerequisite for any change in velocity. This law does not counter the observations we have made via empirical research. The definition of "net external force" will be covered in the section after this one. The force of friction created as an object moves across a flat surface, such as a table or floor, causes the object to slow down and can be attributed to this force (Antwi, 2015). Would there be a reduction in the object's velocity if the force of friction were somehow eliminated?

The concept of cause and effect plays a pivotal role in accurately depicting events occurring in diverse circumstances. For instance, let us contemplate the scenario of an object in motion along a horizontal surface characterized by roughness. The object rapidly comes to a stop. The object exhibits an increased sliding distance when the surface is treated with talcum powder to enhance its smoothness.

The object can achieve a greater sliding distance by applying lubricating oil to the surface due to the increased smoothness. Extending the analysis to a hypothetical scenario involving a frictionless surface, one can envision the object moving in a linear trajectory without any constraints. Friction is the underlying factor responsible for the deceleration, by Newton's first law. If friction were eliminated, the object would experience no deceleration. Let us contemplate an air hockey table. When the airflow is deactivated, the puck exhibits limited displacement as it experiences deceleration due to the opposing force of friction, ultimately coming to a halt.

Nevertheless, the activation of the air results in the generation of an almost frictionless medium, thereby facilitating the puck's prolonged and unhindered gliding over considerable distances without experiencing deceleration. Moreover, with sufficient knowledge regarding the frictional forces at play, it becomes possible to make precise predictions regarding the rate at which the object will decelerate. Friction is considered an external force in the context of classical mechanics (Cohen, 1964). Newton's first law of motion is universally applicable. It can be observed in a wide range of phenomena, encompassing scenarios such as an object sliding on a table, a satellite in orbit, or the circulation of blood propelled by the heart.

Extensive experimentation has provided robust confirmation that alterations in velocity, encompassing changes in speed or direction, necessitate the presence of an external force as the underlying cause. The concept of universally applicable or general laws holds significance within this context and serves as a fundamental characteristic of all laws within the field of physics. Identifying these laws can be likened to recognizing patterns in nature, which subsequently enables the discovery of additional patterns. The intellectual prowess exhibited by Galileo, who initially conceived the notion of the first law, and Newton, who subsequently elucidated it, lay in their inquiry into the fundamental query of causality. The conceptualization of cause and effect as a fundamental framework for understanding phenomena represents a distinct worldview that diverges from the conventional ancient Greek perspective. In the context of the ancient Greeks, inquiries about the rationale behind certain phenomena, such as the existence of stripes on a tiger, would have been addressed through an Aristotelian lens, wherein the response would emphasize the inherent characteristics of the creature in question. While it may be true, this insight lacks practical utility (Erfan & Ratu, 2018).

Inertia is the inherent characteristic of an object to maintain its state of rest or uniform motion. The initial principle formulated by Sir Isaac Newton is commonly referred to as the law of inertia. As evidenced by empirical observations, certain objects exhibit a greater degree of inertia compared to others. As an illustrative instance, altering the trajectory of a sizable boulder is evidently more challenging than that of a basketball.

The measurement of an object's inertia is determined by its mass. Mass can be defined as a quantitative measure of the quantity of substance or matter contained within an object. Determining an object's quantity or amount of matter is contingent upon the numerical representation of atoms and molecules of the diverse classifications it encompasses. In contrast to weight, mass remains constant regardless of location. The mass of an object remains constant regardless of its location, whether it is on Earth, in orbit, or on the surface of the Moon.

In practical applications, quantifying and discerning the entirety of atoms and molecules within an object proves to be exceedingly challenging, thereby rendering the determination of masses through this approach infrequent. In practical terms, the masses of objects are ascertained by comparing them to the standard kilogram (Ding, 2023).

3.4. NEWTON'S SECOND LAW OF MOTION: CONCEPT OF A SYSTEM

The relationship between Newton's second law of motion and Newton's first law of motion is significant. The concept mathematically expresses the causal connection between force and alterations in the state of motion. The second law of motion, formulated by Sir Isaac Newton, is a highly quantitative principle that finds widespread application in analyzing and predicting phenomena involving a force's presence. Before formulating Newton's second law in the form of a concise equation that delineates the precise correlation between force, mass, and acceleration, it is imperative to refine certain concepts that have been previously alluded to (Tawil & Said, 2022).

To begin, it is necessary to grasp the idea of a change in motion before moving on to anything else. One way to summarize the response is as follows: Any change in motion will result in a corresponding change in velocity. According to the definition, the existence of acceleration must be assumed whenever there is a change in the velocity. The presence of a net external force will cause a change in the velocity of an object when Newton's first law is satisfied, as stated in the previous sentence. Consequently, it is clear that an overall external force is responsible for the acceleration that is measured. Almost immediately, another inquiry appears. What exactly does it mean to say that something is an external force? When properly conceptualized, the idea of externality refers to the presence of a force acting from a location physically distinct from the system being considered. The youngster riding in the wagon is an integral part of the analysis's focal point, as shown in Figure 3.4(a), which shows this concept graphically. The other children's actions can be broken down into two separate forces, which make up the exterior pressures they exert. Within the system, a force of nature is a driving force between the components that make up the whole. Looking back at Figure 3.4(a), one realizes that the force that the youngster in the wagon is using to keep their hold on the wagon is an internal force that occurs within the many components that make up the system being considered. The first law of Newton states that the only thing that can alter the motion of a system is the forces acting on it from the outside. There are times when the distinguishability of a system is crystal clear, but there are also moments when the delineation of its boundaries is more grayscale. In many subfields of physics, the fundamental nature of systems and the critical significance of precisely applying Newton's laws have received a lot of attention in recent years. During our investigation into the topic of physics, we will return to the idea that was just mentioned several times (Kulworawanichpong, 2015).

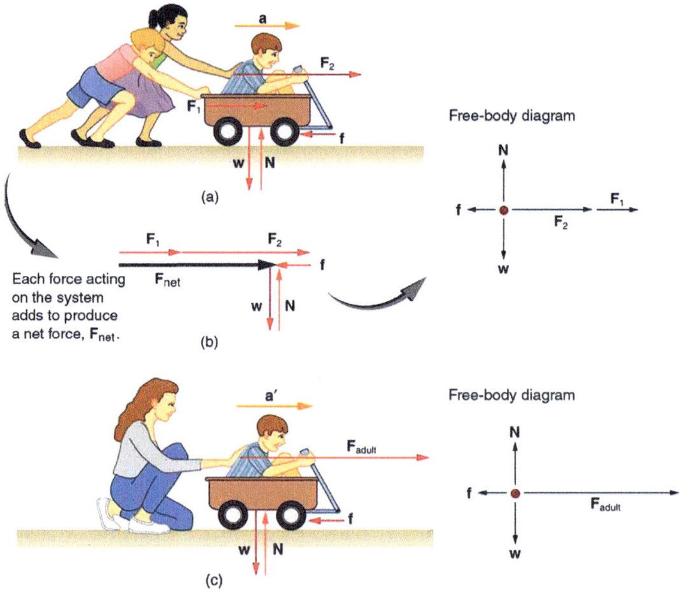

Figure 3.4. Different forces exerted on the same mass produce different accelerations (Source: Openstax, Creative Commons License).

It is logical to posit that acceleration should exhibit a direct proportionality to the net external force acting on a system while aligning with its direction. The assumption above has been empirically validated and is visually depicted in Figure 3.4. In section (a), it is observed that a reduced force results in a correspondingly diminished acceleration compared to the amplified force depicted in section (c). To ensure a comprehensive analysis, the vertical forces are also depicted. It is assumed that these forces nullify each other due to the absence of any vertical acceleration. The vertical forces acting on the object include the weight (w) and the ground's normal force (N) (Zhang et al., 2018).

Additionally, the horizontal force (f) represents the force of friction. The subsequent sections will delve into a more comprehensive analysis of these topics. Currently, friction will be defined as a resistive force that opposes the relative motion between two objects in contact. In Figure 3.4(b), the illustration demonstrates the summation of vector representations of external forces, resulting in the emergence of a net force denoted as F net.

Mathematically.

$a \propto F_{net}$,

Proportionality, denoted by the sign, is represented by the symbol F net, which represents the net external force. Both graphical and mathematical approaches can be utilized to ascertain the external net force. When calculating the vector sum of all external forces, the head-to-tail approach represents the calculation graphically. Alternatively, one can use analysis to calculate the net external force by using the components that make up the external forces. The procedures that are utilized for the addition of vectors in this particular setting are consistent with the procedures that are utilized for the addition of other vectors, as is explained in the subject area of Two-Dimensional Kinematics. The idea that acceleration is directly proportionate to the net external force can be expressed through the proportionality that was just discussed. This idea was presented in verbal form earlier on in this paragraph. After picking the system that will be considered, it is extremely important to recognize the forces coming from the outside while ignoring the forces coming from within the system. It is a big simplification to ignore the myriad of internal forces between the components that make up the system. These forces include the muscle forces that are present within the body of the child as well as the interatomic forces that are present within the objects. However, this simplification enables the correct resolution of highly sophisticated issues, albeit with a tiny margin of error arising from the simplification that was just discussed. This is a significant improvement over the previous scenario (Sujarwanto & Putra, 2018).

It stands to reason that acceleration should have an inverse connection with the system's mass, as this would be consistent with logic. Alternately stated, it is possible to observe that the resultant acceleration created by a particular force is reduced when the mass of an item increases, which also increases the object's inertia. Compared to the acceleration produced by applying an equivalent net external force to a basketball, the acceleration produced by applying the same force to an automobile results in a substantially smaller acceleration. This is demonstrated in Figure 3.5. The relationship between these two quantities can be stated as follows:

$a \propto m_1$

where "m" represents the mass of the system. Empirical investigations have demonstrated that the relationship between acceleration and mass is precisely characterized by inverse proportionality. In contrast, the relationship between acceleration and the net external force is precisely characterized by direct proportionality (Nieminen et al., 2010).

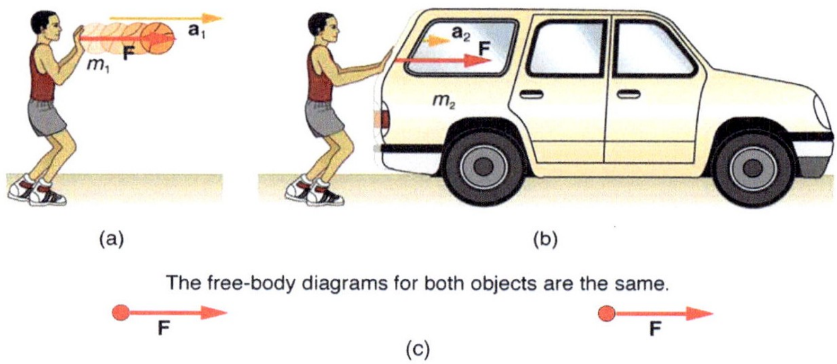

Figure 3.5. The same force exerted on systems of different masses produces different accelerations (Source: Rice University, Creative Commons License).

3.4.1. Newton's Second Law of Motion

A system's acceleration is directly proportional to the net external force operating on the system and in the same direction as that force. On the other hand, the acceleration of a system is inversely proportional to its mass.

Newton's second law of motion can be expressed as an equation as follows:

$$a = mF_{net}.$$

$$F_{net} = ma.$$

Or,

$$F_{net} = ma.$$

While both of these final two equations are essentially equivalent, the first equation provides a deeper understanding of the underlying implications of Newton's second law. The law refers to a causal relationship between three quantities, not solely determined by their respective definitions. The validity of the second law relies entirely on empirical validation (Lemmer, 2018).

3.4.2. Units of Force

In the context of Newton's second law of motion, the relationship between the net force, denoted by F_net, the mass, denoted by m, and the acceleration, denoted by a, is represented by the equation F_net = ma. The objective of this definition is to specify the units of force by referencing the three primary units of measurement: mass, length, and time. The newton (N), part of the International System of Units (SI), is the name given to the magnitude of force necessary to accelerate a system by one meter per second squared when that system has a mass of one kilogram (Wayne, 2012). The definition of the newton may be found in SI's definition of the newton. This is because the total force that is acting on an object is equal to the product of that object's mass and the acceleration that it is experiencing,

$$1\ \text{N} = 1\ \text{kg} \cdot \text{m/s}^2.$$

3.4.3. Weight and the Gravitational Force

When something is allowed to leave its resting position and begins freely falling under the force of gravity, the object experiences an acceleration directed toward the center of the Earth. Newton's second law states that the amount of force exerted upon an object is directly proportionate to the amount of acceleration that the object is experiencing. In the hypothetical situation where the force of air resistance can be disregarded as insignificant, the resultant force acting on a falling object is comparable to the gravitational force, which is more frequently known as the object's weight (abbreviated as w). Because it has a particular directional component, weight, or w, can be represented by a vector. Weight can be thought of as a force that acts in a downward direction since it is aligned with the force of gravity, which also acts in a downward direction. It is usual practice to use the symbol "w" to indicate the amount for the magnitude of weight. Galileo performed a vital role in demonstrating that when air resistance is low, all objects experience a uniform acceleration designated as "g" during free fall. This was one of the most important discoveries that Galileo made. It is feasible to derive an equation that explains the concept of weight by employing the findings of Galileo and applying Newton's second law of motion. This can be done by combining the two (Papaxanthis et al., 1998).

Consider for a moment a material thing with a mass that we'll denote with the letter m and that is now moving in the direction of the Earth in a clockwise direction. The only force the item is subject to is gravity, which exerts itself in a downward direction and has a magnitude indicated by the letter w. Newton's second law states that the external force exerted on an object can be calculated by multiplying the object's mass (m) by its acceleration (a), which results in the equation F net = ma. This is how one can determine the external force being applied to an object.

Given that the object is solely subjected to the gravitational force acting downwards, it can be stated that the net force, denoted as F net, is equal to the object's weight, symbolized as w. It is widely acknowledged that the acceleration experienced by an object as a result of gravitational force is denoted by the symbol "g" or expressed as "a = g" (Podkletnov & Nieminen, 1992).

3.4.4. Weight

Weight or the gravitational force acting on a mass m:

$w = mg.$

Since $g = 9.80\ m/s^2$ on Earth, the weight of a 1.0 kg mass is 9.8 N:

$w = mg = (1.0\ kg) * (9.80\ m/s^2) = 9.8\ N.$

g can assume either a negative or a positive value, contingent upon the direction considered as positive within the coordinate system. It is imperative to consider this aspect when addressing weight-related problem-solving scenarios duly.

The acceleration due to gravity, denoted as g, exhibits slight variations across the Earth's surface, resulting in a location-dependent weight for objects rather than an inherent property. The weight of an object experiences significant variation upon departure from the surface of the Earth. On the lunar surface, specifically on the Moon, the acceleration resulting from the force of gravity is a mere 1.67 m/s^2. A mass of 1.0 kg exhibits a gravitational force of approximately 9.8 N when situated on Earth, whereas its weight is reduced to approximately 1.7 N when placed on the Moon (Liangzao et al., 2010). In this context, the most inclusive interpretation of weight pertains to the gravitational force exerted on an object by the closest substantial celestial body, the Moon, such as Earth, the Sun, and other similar entities. The following definition of weight in physics is widely recognized and holds significant practical value. There is a significant disparity, however, between the definition of weight employed by NASA and the portrayal of weight in popular media regarding space exploration and travel. The terms "weightlessness" and "microgravity" are essentially synonymous with the scientific concept known as "freefall" in the field of physics. The definition above of weight will be employed, and a meticulous differentiation will be made between free-fall and true weightlessness.

It must be acknowledged that mass and weight are distinct physical quantities, albeit closely related. In classical physics, mass measures an object's quantity of matter, representing the amount of "stuff" it contains. It remains constant regardless of the gravitational field. On the other hand, weight refers to the gravitational force exerted on an object and is subject to variation based on the strength of the gravitational field. There is a temptation to conflate the two concepts, given that most of our illustrations are situated on Earth, where the gravitational force exerted on an object exhibits only minor fluctuations in correspondence with the object's position. Moreover, it is common for the phrases mass and weight to be utilized interchangeably in everyday discourse. For instance, in our medical documentation, our "weight" is frequently indicated in kilograms rather than the appropriate units of newtons (Legaye & Duval-Beaupere, 2008).

3.5. NEWTON'S THIRD LAW OF MOTION: SYMMETRY IN FORCES

The musical Man of la Mancha features a segment relevant to Newton's third law of motion. While narrating a domestic dispute to Don Quixote, Sancho affirms, "Truly, I responded in kind, Your Grace. However, it is pertinent to acknowledge that she exhibits a higher level of gritthan my own." According to the proverb, "Regardless of whether the stone hits the pitcher or the pitcher hits the stone, the outcome will be disadvantageous for the pitcher." The phenomenon above arises when an object exerts a force upon another object, leading to the former object experiencing a force of equal magnitude but opposite direction. This claim is supported by numerous everyday instances, such as accidentally colliding one's toe with a stationary object or projecting a spherical object through the atmosphere. The explicit statement of Newton's third law of motion is as follows. When an object exerts a force on another object, the first object experiences a force equal in magnitude but opposite in direction to the force it exerts (Wimmer et al., 2013).

The legislation in question encapsulates a fundamental symmetry principle commonly observed in the natural world. This principle states that forces always occur in pairs, and when one entity exerts a force on another, it inevitably experiences a corresponding force in return. The law under examination is commonly known in informal language as "action-reaction." In this context, the force exerted is identified as the action, while the force experienced in response is labeled as the reaction. The practical implications of Newton's third law pertain to examining the sources of force and the determination of external forces exerted on a given system.

The observation of human locomotion clearly demonstrates the manifestation of Newton's third law. The depicted scenario in Figure 3.6 entails a swimmer commencing propulsion by pushing off from the perimeter of a swimming pool. The person exerts a force on the pool wall by utilizing her feet, leading to an acceleration that opposes the direction of the applied force.

The swimmer has encountered a reactive force equal in magnitude but opposite in direction to that exerted by the wall. It is a commonly held assumption that two opposing forces of equal magnitude would cancel each other out. However, this assumption does not hold when considering their impact on separate systems.

Within this specific scenario, two systems merit further investigation: the swimmer and the wall. When considering the swimmer as the system of interest, as illustrated in the diagram, the wall exerts an external force on the system, referred to as F wall on feet, which impacts the swimmer's movement. The swimmer propels themselves towards the wall labeled as "F" by utilizing their lower extremities (Yifat et al., 2018).

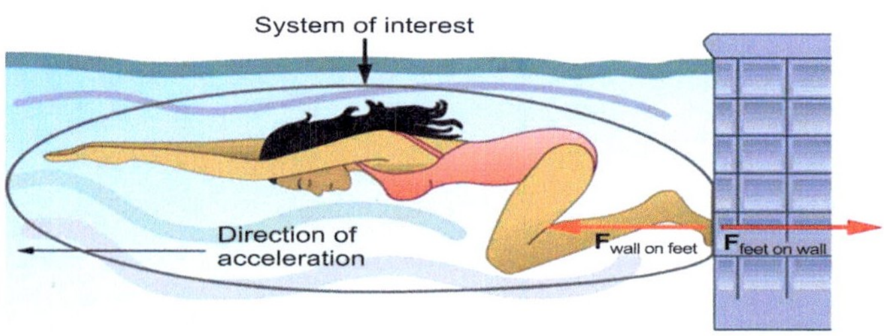

Figure 3.6. When the swimmer exerts a force F feet on the wall on the wall, she accelerates in the direction opposite to that of her push. This means the net external force on her is in the direction opposite to F feet on the wall (Source: Rice University, Creative Commons License).

In contrast, the force labeled as F applied to the wall is oriented towards the wall itself rather than our particular system under examination. Hence, the force applied by the feet against the wall does not directly influence the system's motion and does not oppose the force exerted by the wall on the feet. Acknowledging that the swimmer applies force in a direction that opposes her intended motion is important. As a result, the reaction to her effort is the desired result.

Numerous additional instances exemplify Newton's third law. As an academic instructor moves deliberately across the space in front of a whiteboard, she applies a force in the opposite direction to the floor. The professor experiences a reaction force exerted by the floor in the forward direction, resulting in her acceleration forward. In a similar vein, the acceleration of a car occurs as a result of the reactive force exerted by the ground on the drive wheels, which is in response to the drive wheels exerting a backward force on the ground. The phenomenon of wheels exerting a backward force can be observed when tires rotate on a gravel surface, resulting in the projection of rocks in the opposite direction. In an alternative illustration, rockets propel themselves in a forward direction through gas expulsion in the opposite, rearward direction at a considerable velocity. This implies that the rocket applies a substantial rearward force on the gas within its combustion chamber, resulting in a significant forward reaction force exerted by the gas on the rocket. The force that arises due to a reaction is called thrust. A prevalent misconception exists that rockets achieve propulsion by exerting force against the ground or the air in their rearward direction. Their performance is enhanced in a vacuum environment, facilitating the efficient expulsion of exhaust gases. Helicopters generate lift through the downward displacement of air, applying an upwards reaction force. Both birds and airplanes achieve flight by applying a force to the surrounding air in a direction that opposes the force required for their movement. For instance, the avian wings generate a downward and backward airflow, thereby generating lift and facilitating forward motion. The locomotion of an octopus in aquatic environments involves the expulsion of water through a specialized funnel located on its body, akin to the propulsion mechanism employed by a jet ski. Like Sancho's circumstance, professional practitioners of cage fighting encounter reaction forces upon delivering punches, occasionally resulting in hand fractures upon striking an adversary's physique (Bayraktar et al., 2010).

3.6. NORMAL, TENSION, AND OTHER EXAMPLES OF FORCES

Various terms refer to forces, including pulling, pushing, lifting, weight, thrust, friction, and tension. Historically, forces have been classified into various categories and assigned nomenclature based on their origin, mode of transmission, or resultant impacts. This section will discuss the most significant categories, along with notable applications. Additional illustrations of forces will be examined in subsequent sections of this discourse (Malm et al., 2013).

3.6.1. Normal Force

Weight, also referred to as the force of gravity, is an omnipresent force that operates continuously and necessitates counteraction to prevent an object from descending. It is readily apparent that the necessity to exert an upward force to maintain a substantial object's immobility is evident, as depicted in Figure 3.7(a). However, the mechanism by which inanimate objects, such as a table, can bear the weight of a mass placed upon them, as depicted in Figure 3.7(b), remains a subject of inquiry. When a dog food container is positioned on the table's surface, the surface exhibits a slight deformation due to the applied load. Observing load placement on a card table would result in noticeable effects.

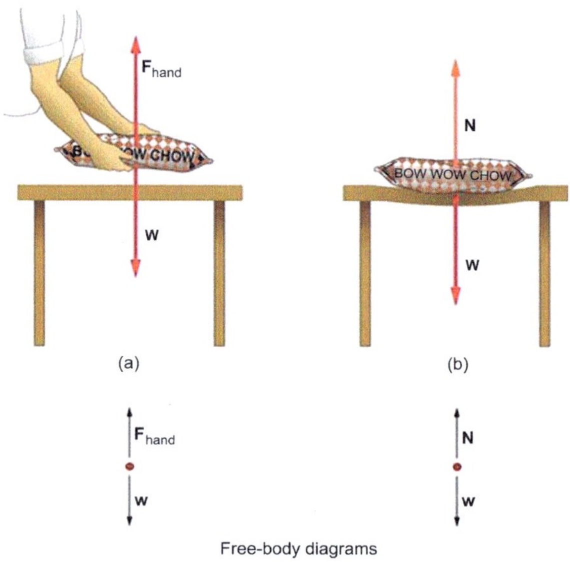

Figure 3.7. (a) The person holding the bag of dog food must supply an upward force F hand equal in magnitude and opposite in direction to the weight of the food w. (b) The card table sags when the dog food is placed on it, much like a stiff trampoline (Source: Openstax, Creative Commons License).

However, it is important to note that even objects with rigidity experience deformation when subjected to external forces. Unless the object undergoes deformation beyond its elastic limit, it will generate a restoring force analogous to a deformed spring. There is a direct relationship between the magnitude of deformation and the magnitude of the restoring force. When an external force is applied to the table, it causes it to deform, resulting in a downward sag. This deformation continues until the restoring force exerted by the table reaches an equilibrium with the weight of the load. Currently, the total external force acting on the load is in equilibrium. This scenario pertains to a stationary load placed on the table. The table exhibits rapid sagging, albeit negligible, that may go unnoticed. However, it resembles the phenomenon observed in trampolines where the surface sags upon weight exertion (Wang & Lu, 2023).

3.6.2. Tension

Tension is defined as the application of a force in the longitudinal direction of a medium, specifically when transmitted through a flexible medium such as a rope or cable. The term "tension" has its etymological roots in a Latin term that denotes the action of stretching. The anatomical structures referred to as tendons are aptly named due to their transmitting muscle forces to different body regions. A pliable linkage, such as a thread, rope, cord, wire, or cable, can apply forces exclusively in the direction that aligns with its length. Therefore, the transmission of force through a flexible connector is identified as tension, whereby the direction of the force is aligned parallel to the connector. It is crucial to understand that tension denotes the action of applying a pulling force within a connector. Alternatively, let us consider the statement: "The act of pushing a rope does not result in the application of force." The tension force manifests as an outward pulling effect on both extremities of a rope (Shinogaya et al., 2001).

The individual depicted in Figure 3.8 is observed to be grasping a mass suspended by a rope.

The equilibrium condition of a rope can be established by demonstrating that the tension in the rope is equivalent to the weight of the mass it supports, as substantiated by Newton's second law. If the mass with a weight of 5.00 kg depicted in the diagram is in a state of rest, then its acceleration is determined to be zero, concluding that the net force acting on it is also zero. The sole external forces exerted on the mass consist of its weight, denoted as w, and the tension, represented as T, provided by the rope. Thus,

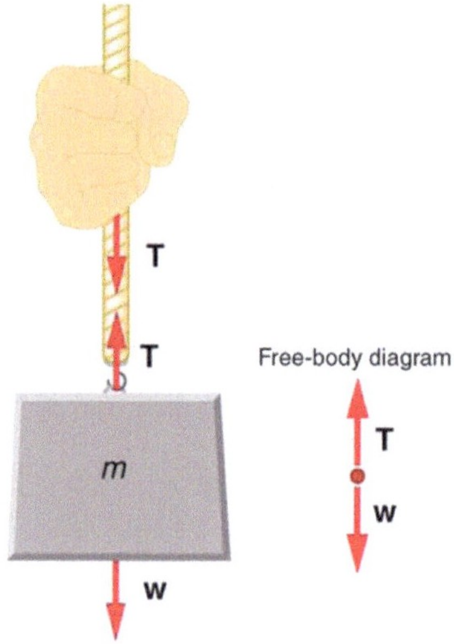

Figure 3.8. When a perfectly flexible connector (one requiring no force to bend it) such as this rope transmits a force T, that force must be parallel to the length of the rope, as shown. The pull such a flexible connector exerts is tension (Source: Openstax, Creative Commons License).

F net = T − w = 0,

The variables T and w represent the magnitudes of tension and weight, respectively. The signs of these variables indicate their respective directions, with the positive direction being upward in this context. Accordingly, as anticipated, the magnitude of the tension force is equivalent to the gravitational force exerted by the mass being supported (Hejazi et al., 2013):

$T = w = mg.$

For a 5.00-kg mass,

T = mg = 49.0 N

By severing the rope and introducing a spring, it is possible to witness and quantify the tension force exerted on the rope through the elongation of the spring, which would correspond to a force magnitude of 49.0 N. Flexible connectors are frequently employed in various applications to facilitate the transmission of forces around curved paths. Such applications include hospital traction systems, finger joints, and bicycle brake cables. Without friction, the tension remains unaltered as it is transmitted. The sole alteration is in its orientation, which remains parallel to the pliable linkage. This phenomenon is depicted in Figure 3.9(a) and (b) (Zammit et al., 2010).

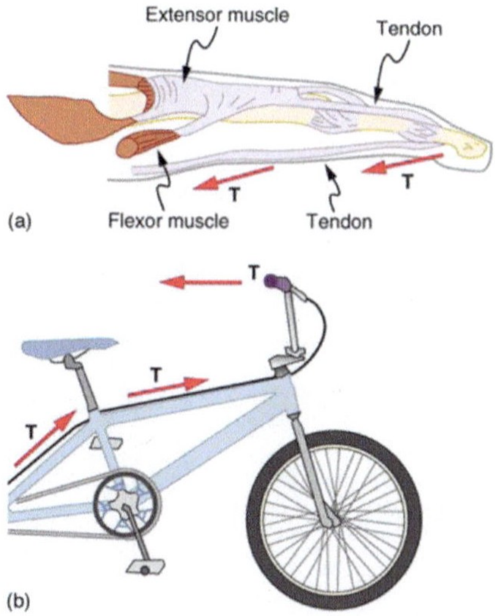

Figure 3.9. The tendons in the fingers transmit force, denoted as T, from the muscles to various regions of the finger. These tendons exhibit the characteristic of frequently altering the direction of force transmission while maintaining a consistent magnitude (Source: Rice University, Creative Commons License).

3.6.3. Extended Topic: Real Forces and Inertial Frames

In addition to the types above, an additional distinction exists among forces. Certain forces possess objective existence, while others lack such reality. Real forces are classified as those that arise from a tangible source, such as the force of gravity. In contrast, fictitious forces emerge solely due to the presence of an observer within an accelerating frame of reference, such as a rotating system (e.g., a merry-go-round) or a system experiencing linear acceleration (e.g., a decelerating car). As an illustration, when a satellite traverses in a northerly direction over Earth's northern hemisphere, it will manifest an apparent westward force to an observer on Earth, which lacks a tangible source. The phenomenon observed in this scenario is the Earth's eastward rotation, causing the satellite to move in an easterly direction. From Earth's reference frame perspective, the observed phenomenon can be characterized as a force acting in the westward direction on the satellite (de Pereira et al., 2016).

All the forces examined in this section are indeed genuine forces. However, it is worth noting that additional real forces exist, such as lift and thrust, which have not been addressed within the confines of this particular section. They exhibit a higher degree of specialization, rendering it unnecessary to examine every force category comprehensively. It is a common inquiry to question the fundamental simplicity that we strive to uncover within the extensive array of forces in the field of physics. Do certain concepts possess a higher level of complexity compared to others? Do certain phenomena exhibit distinct manifestations despite being driven by a common underlying force? Both questions can be answered affirmatively, as demonstrated in the subsequent section, and further elaborated upon in the later discussion on modern physics within this text (Czajko, 2022).

3.7. FURTHER APPLICATIONS OF NEWTON'S LAWS OF MOTION

This section presents several additional intriguing applications of Newton's laws of motion. Furthermore, these examples also serve to elucidate additional intricacies within the field of physics and contribute to developing problem-solving abilities (White, 1984).

The bathroom scale is a noteworthy illustration of the normal force exerted on an object. The given measurement offers a numerical assessment of the force required to counteract the downward gravitational force and maintain equilibrium for a given object. However, is it possible to anticipate the reading displayed on the dial of a bathroom scale when an individual stands on it while inside an elevator in motion? Is it possible to observe a numerical value that exceeds one's body weight upon initiating elevator movement? In the scenario where the elevator ascends at a consistent velocity, does the scale continue to indicate a weight greater than that observed at rest?

Physics becomes particularly captivating and influential when employed in broad contexts encompassing a wide range of physical principles rather than being limited to a narrow set of circumstances. Integrating Newton's laws of motion with previously discussed concepts in this text enables the resolution of motion-related problems. An illustration of the relationship between forces and accelerations can be found within the realm of kinematics, thereby establishing the significance of the preceding chapters. When confronted with problems that encompass a multitude of forces, velocity, acceleration, and position, it is advisable to employ the subsequent steps (a strategy for problem solving) to tackle the problem: The first step entails the identification of the physical principles that are relevant to the situation at hand. Identifying the principles involved can be facilitated by listing the given information and the quantities to be calculated. Proceed to solve the problem by employing the strategies delineated in the text. If these resources are accessible and relevant to the given subject matter, consulting and citing them is advisable. It is advisable to refer to the relevant sections of the text that pertain to a specific topic. The subsequent exemplification demonstrates the application of these strategies to an integrated concept problem (Pospíšil, 2013).

3.8. THE FOUR BASIC FORCES

One of the most notable simplifications in physics is that a mere four distinct forces are sufficient to explain and encompass all presently understood phenomena. Indeed, it is noteworthy that the vast majority of the forces we encounter firsthand can be attributed to a singular fundamental force, the electromagnetic force. The gravitational force is the sole force humans directly encounter, which does not fall under electromagnetic forces. This oversimplification greatly underestimates the vast array of distinct forces, as only a limited number of them were addressed in the preceding section. It is widely understood that the fundamental forces operate using the transmission of subatomic mediator particles, and the properties of these forces are contingent upon the nature of the particles involved in the exchange process. The phenomenon of action at a distance, exemplified by the gravitational force exerted by Earth on the Moon, is elucidated through the concept of a force field rather than through direct physical contact (Goldie et al., 1989).

The fundamental forces encompassing the physical interactions in nature consist of the gravitational force, the weak nuclear force, the electromagnetic force, and the strong nuclear force. The properties of the subject under consideration are briefly presented in Table 3.1. Due to their limited range, spanning the size of a nucleus or smaller, the strong and weak nuclear forces are not directly perceptible to us. However, their significance lies in their fundamental role in the composition of matter. These forces are responsible for determining the stability of nuclei and their propensity to undergo decay. Additionally, they serve as the fundamental mechanisms underlying energy discharge in specific nuclear reactions. The stability of atomic nuclei and the relative distribution of elements in nature are both influenced by nuclear forces. The characteristics of an atom's nucleus are responsible for determining its electron count, consequently indirectly influencing the atom's chemistry. Further discussion on these subjects will be provided in subsequent chapters.

The exposition will delve into a more comprehensive examination of the four fundamental forces as you advance further in the discourse. The concept of gravitational force is expounded upon in the context of Uniform Circular Motion and Gravitation. In contrast, the principles governing electric force are discussed in Electric Charge and Electric Field. Similarly, the phenomenon of magnetic force is elucidated in the domain of Magnetism, and the intricacies of nuclear forces are explored in the realm of Radioactivity and Nuclear Physics. At the macroscopic level, electromagnetism and gravity are the fundamental forces underlying all other forces. Nuclear forces play a crucial role in the fundamental structure of matter, yet their effects are not directly observable at the macroscopic level (Charnes et al., 1984).

Table 3.1. Properties of the Four Basic Forces (Source: Rice University, Creative Commons License)

Force	**Approximate Relative Strengths**	**Range**	**Attraction/Repulsion**	**Carrier Particle**
Gravitational Force	10 to 38 times	∞	Only attractive	Graviton
Electromagnetic Force	10 to 2 times	∞	Both attractive and repulsive	Photon
Weak Nuclear Force	10 to 13 times	<10 to 18 m	Both	W+;W–;Z 0
Strong Nuclear Force	1 time	<10 to 15 m	Both	Gluons

The gravitational force exhibits a relatively low magnitude, which can be attributed to gravity, and is inherently attractive, thereby enabling its perceptibility. An individual's weight is determined by the gravitational force exerted by the Earth. At the macroscopic level, such as in astronomical systems, the gravitational force plays a paramount role in dictating the movements of celestial bodies such as moons, planets, stars, and galaxies. The gravitational force influences the fundamental properties of space and time. As will be demonstrated in a subsequent examination of general relativity, it is observed that the spatial dimension exhibits curvature in the proximity of highly massive entities, such as the Sun. Additionally, temporal progression experiences a deceleration near such massive entities (Liu et al., 2007). Electromagnetic forces can exhibit either attractive or repellent behavior. Long-range forces are characterized by their ability to act over vast distances, and their effects on macroscopic objects are nearly nullified. Remembering that the net external force holds significance in this context is crucial. If the cancellation did not occur, it would be observed that electromagnetic forces would exert a dominant influence over the gravitational force. The electromagnetic force is a composite of electrical forces, encompassing phenomena like static electricity and magnetic forces, which influence the behavior of a compass needle. The distinction between these two forces was widely accepted until the early 19th century, when researchers realized that they are, in fact, different manifestations of a single force. The discovery above exemplifies a paradigmatic instance of force unification.

Comparably, the forces of friction, tension, and other force categories that we encounter firsthand (excluding gravity) arise from the electromagnetic interactions between atoms and molecules. In specific applications, it remains advantageous to analyze these forces individually due to the distinct manners in which they become apparent.

Contemporary physicists are currently engaged in the investigation of potential interconnections among the four fundamental forces. Efforts to consolidate all fundamental forces into a single framework fall within the scope of Grand Unified Theories (GUTs), which have recently demonstrated certain degrees of accomplishment. It has been established that in circumstances characterized by exceedingly high density and temperatures, such as those prevailing in the early stages of the universe, the electromagnetic and weak nuclear forces exhibit indistinguishable characteristics.

The electroweak force can now be regarded as encompassing distinct manifestations. The original set of four items has been effectively reduced to a set of three.

Achieving comprehensive unification of all forces presents significant challenges, particularly regarding incorporating the gravitational force. This force possesses unique attributes that influence the fabric of space and time within which the other forces operate (Moody et al., 2003).

Although the unification of forces does not have any immediate implications for the discussion of forces in this text, it is intriguing to note the presence of underlying simplicity amidst the apparent complexity of the cosmos. The simplicity of nature is not contingent upon any specific rationale; rather, it is an inherent characteristic.

All forces exert their influence over objects from a certain distance. The gravitational force is apparent. For instance, the Earth and the Moon engage in interactions without physical contact. This statement holds across all other forces as well. Friction, exemplified by electromagnetic force among atoms that do not make physical contact, is a significant factor to consider. What is the medium through which forces are transmitted between objects? One potential approach to addressing this inquiry involves conceptualizing the presence of a force field enveloping the entity responsible for generating the force.

The presence of a second object, commonly referred to as a test object, within this field, will result in the exertion of a force, which is dependent on the object's position and various other factors. In this context, the field serves as the medium through which force is transmitted between objects. The field is defined as an inherent property of the object that generates it, and it remains unaffected by the presence of a test object within it. Earth's gravitational field is determined by two factors: the mass of Earth and the distance from its center.

This gravitational field remains constant regardless of the presence or absence of other masses. Utilizing the concept of a field proves advantageous due to its ability to facilitate the formulation of equations that describe force fields encompassing objects. For instance, in the case of gravity, this approach yields the equation w = mg when considering Earth's surface. Consequently, these equations enable the computation of motions associated with such force fields (Figure 3.10) (Elo et al., 2015).

The concept of fields has been effectively employed, yielding notable achievements in scientific inquiry. We can accurately compute motions and provide comprehensive descriptions of natural phenomena using field equations. Despite its usefulness, the field concept fails to address the issue of the entity responsible for exerting force.

There has been a proposal in the past few decades, specifically since 1935. The graviton is a hypothesized elementary particle that remains unverified through empirical observation conducted by the scientific community. Please refer to the subsequent section for an in-depth analysis of gravitational waves. The particles W+, W-, and Z0 are vector bosons. Theoretical frameworks initially postulated these particular particles and detected them for the first time in 1983. Scientists have proposed eight distinct types of gluons, which are indicated by the exchange of mesons within atomic nuclei.

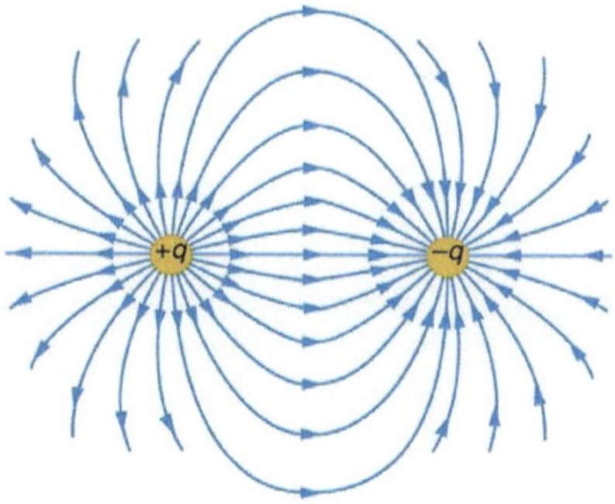

Figure 3.10. The electric force field between a positively and negatively charged particle. When a positive test charge is placed in the field, the charge will experience a force in the direction of the force field lines (Source: Rice University, Creative Commons License).

Hideki Yukawa (1907–1981) made significant contributions to the understanding of the strong nuclear force, proposing that the transmission of all forces occurs through the exchange of elementary particles. Particle exchange can be conceptualized as a parallel to macroscopic events, where two individuals engage in a reciprocal basketball exchange, exerting a repulsive force between them without direct physical contact (Figure 3.11) (Birch & Dickinson, 2003).

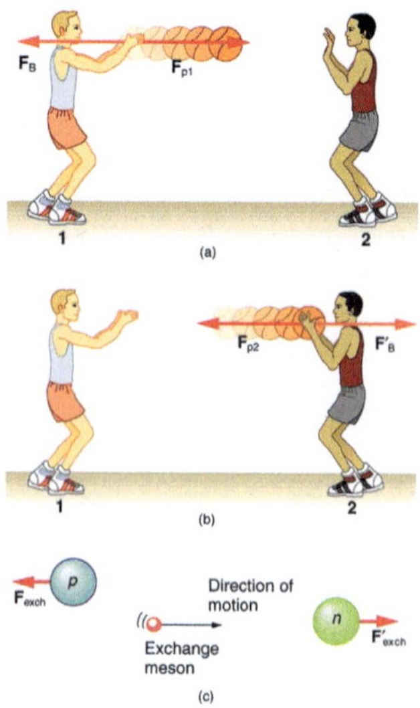

Figure 3.11. The exchange of masses resulting in repulsive forces (Source: Openstax, Creative Commons License).

The notion of particle exchange enhances, rather than contradicts, the fundamental principles of field theory. From a philosophical standpoint, the notion of a physical entity actively traversing the space between objects that exert influence over each other from a distance is deemed more intellectually gratifying. Table 3.1 provides a comprehensive compilation, encompassing the exchange or carrier particles associated with the four fundamental forces, including both observed and proposed particles. However, the significant outcome of the particle-exchange proposition lies in the discovery of Yukawa's hypothesized particle, alongside several other unforeseen particles, which subsequently sparked further investigations. The culmination of extensive research efforts ultimately resulted in the proposition of quarks as the fundamental constituents of matter, a fundamental principle within the Grand Unified Theories (GUTs) framework. If these theories prove to be successful, they will provide explanations for forces and the fundamental structure of matter. However, it is important to note that physics is fundamentally an empirical discipline requiring experimental validation to test the validity of these theories within the realm of the physical world. Currently, researchers at the CERN laboratory in Switzerland are commencing experimentation on these hypotheses using the Large Hadron Collider, the largest particle accelerator globally. The accelerator, with a circumference of 27 kilometers, facilitates the collision of two high-energy proton beams, which are propelled in opposite directions. A total energy of 14 million electron volts will be accessible. There is an expectation that certain novel particles, potentially serving as force carriers, may be discovered. The Higgs boson is considered one of the force carriers researchers eagerly seek to detect. The examination of the properties of particles may provide insights into the underlying factors contributing to the variations in their respective masses (Figure 3.12) (Israelachvili & McGuiggan, 1990).

Figure 3.12. The world's largest particle accelerator spans the border between Switzerland and France. Two beams, traveling in opposite directions close to the speed of light, collide in a tube similar to the central tube shown here (Source: Frank Hommes, Creative Commons License).

The wave-like behavior of tiny particles will be further examined in a subsequent chapter. To gain a more comprehensive understanding of force-carrier particles, it is instructive to examine the phenomenon of gravity from an alternative standpoint. The pursuit of gravitational waves has been underway for a considerable time. An apparatus for detection has been constructed in the United States, comprising two substantial installations located approximately 3000 km apart. These installations are situated in Washington State and Louisiana. The establishment is formally called the Laser Interferometer Gravitational-Wave Observatory (LIGO). Every installation is specifically engineered to utilize optical lasers to scrutinize even the slightest displacement in the relative positions of two masses caused by the influence of gravitational waves. The two locations enable the concurrent assessment of these minute phenomena to be distinguished from other naturally occurring events, such as seismic activities. The detectors were first put into operation in 2002, and efforts are underway to enhance their sensitivity. Italy, Germany, and Japan have constructed comparable installations, namely VIRGO, GEO600, and TAMA300,to establish a global network of gravitational wave detectors (Hendricks, 1941). The realm of international cooperation in this domain is expanding its scope to encompass outer space through the collaborative efforts of the European Union and the United States in the Laser Interferometer Space Antenna (LISA) project. Monitoring spacecraft will effectively mitigate the impact of earthquakes and other terrestrial disturbances.

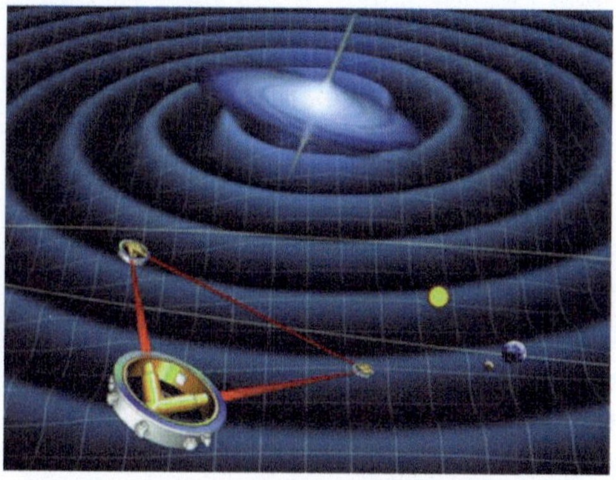

Figure 3.13. Future experiments for measuring gravitational waves in a space-based setting. Presented in this illustration is a depiction of the orbit of the Laser Interferometer Space Antenna (LISA) (Source: NASA, Creative Commons License).

The Large Interferometer Space Antenna (LISA) will serve as a valuable complement to the Laser Interferometer Gravitational-Wave Observatory (LIGO) by focusing on studying significantly more massive black holes. This will be achieved by observing gravitational-wave sources that emit substantially longer wavelengths. Three satellites will be deployed in outer space, positioned in an equilateral triangle configuration with sides measuring 5,000,000 kilometers each. The proposed system aims to ascertain individual satellites' relative positions to detect the presence of gravitational waves as they pass through. To detect any waves, a level of accuracy within a range of 10% of the size of an atom will be required. The initiation of this project has the potential to occur as early as the year 2018 (Figure 3.13) (De Oliveira et al., 2012).

3.9. FRICTION AND DRAG FORCES

Frictional and drag forces are fundamental phenomena in physics that oppose objects' motion as they traverse different mediums. Although both forces are related to opposing motion, they exhibit distinctions in their fundamental principles and practical implementations. This section will delve into the theoretical frameworks of kinetic and static friction, the coefficients of friction, various forms of drag forces (with particular emphasis on air resistance), and strategies for mitigating friction and drag (Gibilaro et al., 1985).

3.9.1. Static vs. Kinetic Friction

Friction is the physical phenomenon characterized by the resistance encountered when two surfaces in contact attempt to move or are in relative motion. The phenomenon emerges due to the atomic or molecular scale interactions between the surfaces involved. There exist two primary categories of friction, namely static friction and kinetic friction (Hashiguchi & Ozaki, 2008). The mathematical expression for the force of static friction can be derived as follows:

$$F_{\text{static}} \leq \mu_s \cdot N$$

The symbol Fstatic represents the force of static friction. At the same time,μs denotes the coefficient of static friction, a dimensionless constant that relies on the specific characteristics of the surfaces in contact. Additionally, N represents the normal force, which is the force exerted by a supporting surface to counteract the object's weight. Kinetic friction refers to the resistive force opposing the relative motion between two sliding surfaces. Once an object surpasses the threshold of static friction and initiates motion, the force of kinetic friction becomes operative.

The mathematical expression for the force of kinetic friction can be derived as follows:

$$F_{\text{kinetic}} = \mu_k \cdot N$$

The friction coefficientsabout a specific combination of materials are ascertained through empirical investigation. They are subject to variation based on factors such as surface roughness, temperature, and the application of lubricants between the contacting surfaces (Kotwal & Bhushan, 1996).

3.9.1.1. Coefficients of Friction

The dimensionless constants μs and μk represent the values of the coefficients of friction, which quantitatively describe the frictional force generated by two surfaces in contact. The dissimilarities between static and kinetic friction arise from their distinct characteristics.

The coefficient of static friction, denoted as μs, measures the maximum amount of frictional force that can be applied between two surfaces at rest without initiating any relative motion between them. The coefficient of static friction is generally higher than that of kinetic friction.

The coefficient of kinetic friction, denoted as μk, measures the frictional force that arises between two surfaces when they are in motion relative to each other (Sanders et al., 1998).

The friction coefficients may exhibit considerable variation depending on the material characteristics of the contacting surfaces. An illustration of this phenomenon can be observed in the dissimilarity between the coefficient of friction exhibited by rubber on dry pavement and that displayed by ice on steel. Engineers and designers consider these coefficients when designing systems to ensure appropriate functionality and safety (Chowdhury & Helali, 2008).

3.9.2. Types of Drag Forces

The phenomenon of drag force, alternatively referred to as air friction or fluid opposition, represents a form of resistance that opposes an object's movement through a fluid medium, such as air or water. Drag forces are significant in multiple disciplines, such as aerodynamics, fluid mechanics, and transportation.

The determination of drag force is contingent upon various factors, encompassing the morphology, dimensions, velocity, and fluid characteristics relevant to the object in motion. There exist primarily two distinct categories of drag forces (Hsu et al., 2016):

Pressure drag, also known as form drag, is a phenomenon that occurs when an object moving through a fluid experiences resistance due to the pressure difference between the front and back of the object. Pressure drag is a phenomenon that arises due to the disparity in pressure between the leading and trailing surfaces of an object in motion through a fluid medium. When an object undergoes motion within a fluid medium, the fluid particles experience compression in the region preceding the object, generating a localized zone characterized by elevated pressure. The presence of an object induces a phenomenon where fluid particles tend to occupy the space generated by the said object, resulting in the formation of a region characterized by reduced pressure. The disparity in pressure generates a resultant force that opposes the object's motion, impeding its displacement.

The object's geometric configuration and frontal cross-sectional area significantly influence the pressure drag's magnitude. Objects that possess streamlined and aerodynamic shapes tend to encounter lower levels of pressure drag compared to flat objects or irregular shapes (Lubarda & Lubarda, 2022).

Skin friction, also referred to as viscous drag, arises due to the frictional resistance between an object's surface and the fluid's particles. As the object traverses the fluid medium, the fluid particles in immediate proximity to the object's surface undergo a reduction in velocity compared to the surrounding freestream fluid, thereby inducing shear stress at the object's surface. The shear stress experienced by the object gives rise to a drag force that acts in the opposite direction to its motion.

The skin friction experienced by an object is contingent upon the surface characteristics of the object, specifically its smoothness or roughness, as well as the viscosity of the fluid in which it is immersed. Smoother surfaces and fluids with lower viscosity exhibit reduced levels of skin friction.

In numerous practical scenarios, the combined effects of pressure drag and skin friction collectively contribute to the total drag force encountered by an object in motion within a fluid medium (Morris et al., 2004).

3.9.3. Reducing Friction and Drag

Reducing friction and drag plays a pivotal role in diverse engineering applications, as it enhances efficiency, conserves energy, and optimizes performance. The following are several commonly utilized techniques for reducing frictional forces and drag forces:

1. Lubrication involves the application of substances, such as oily substances and greases, to minimize the friction experienced by surfaces in contact. Lubrication reduces frictional forces by interposing a slippery layer on both surfaces, thus minimizing direct contact.
2. Streamlining, a concept employed in aerodynamics and transportation, entails deliberately designing objects with sleek and streamlined shapes to minimize the effects of pressure drag. Streamlining is a technique that enhances the efficiency of fluid movement by reducing sudden shifts in shape and minimizing frontal area (Abdulbari et al., 2018).
3. Surface treatment techniques such as smoothing and polishing can decrease skin friction by reducing surface roughness, which in turn reduces the shear stress experienced by the fluid. In certain instances, the implementation of dimples or texture on the surface can be employed to regulate the flow of the boundary layer and mitigate drag.
4. Active control systems are utilized in various applications, including modern aircraft and vehicles, to effectively manipulate surfaces or components to optimize the flow surrounding them. These systems exhibit a continuous capacity to adapt to dynamic conditions, resulting in a reduction of drag and an enhancement of performance (Gee et al., 2015).
5. Operating at the optimum velocity has the potential to decrease drag forces in certain circumstances. One illustrative instance involves the potential for vehicles to attain enhanced fuel efficiency at specific velocities, owing to the equilibrium established between engine torque and drag resistance (Hashima et al., 2015).

3.10. WORK, ENERGY, AND POWER

The concepts of energy, work, and power are fundamental principles in physics, serving as essential components in comprehending the dynamics of physical systems. The concepts above exhibit a strong correlation and interdependence, serving as the fundamental framework for examining mechanical operations and transformations. This section will delve into the principles of the work-energy theorem, encompassing the domains of potential and kinetic energy, the preservation of mechanical energy, and the fundamental notions of power and efficiency (Perrin et al., 1987).

3.10.1. Work-Energy Principle

The work-energy principle is an elementary concept in mechanics. It establishes a relationship between the work exerted on an object and the resulting change in its kinetic energy. According to the work-energy principle, the amount of work performed on an object is equivalent to the alteration in its kinetic energy.

Tip

Keep in mind that work done on a system is equal to the change in its kinetic energy, according to the work-energy principle.

In a mathematical context, the work-energy principle can be formally articulated as follows:

$$W=\Delta KE$$

where W represents the amount of work exerted on the object, while ΔKE denotes the object's kinetic energy alteration.

In mechanics, work is defined as multiplying the force exerted on an object by the change in the object's position in the same direction as the force. The quantity in question is a scalar, and its unit of measurement is the joule (J).

Kinetic energy, denoted as KE, refers to the energy an object possesses due to its motion. The relationship between the mass (m) of an object and the square of its velocity (v) is directly proportional:

$$KE=\frac{1}{2}mv^2$$

The work-energy principle elucidates the correlation between work and kinetic energy, highlighting that work transfers energy to or from an object, thereby inducing a modification in its motion (Tsai, 2015).

3.10.2. Kinetic Energy and Potential Energy

Potential and kinetic energy are two distinct forms of mechanical energy that an object can possess. Mechanical energy refers to the energy inherently linked to an object's position and motion (Schwendeman, 1966).

3.10.2.1. Kinetic Energy (KE)

As previously stated, kinetic energy refers to the energy that an object possesses due to its motion. The outcome is contingent upon the mass of the object and its velocity. The formula can be utilized to compute the kinetic energy of an object:

$$KE = \frac{1}{2} mv^2$$

The relationship between the kinetic energy of an object and its mass is directly proportional, while the kinetic energy also increases exponentially with the square of its velocity. This phenomenon can be observed whereby the doubling of velocity leads to a quadrupling of kinetic energy (Thompson, 1980).

3.10.2.2. Potential Energy (PE)

Potential energy refers to the energy that an object exhibits due to its specific position or arrangement about a designated reference point. Multiple manifestations of potential energy exist, with two prevalent categories being gravitational and elastic potential energy.

Gravitational potential energy (GPE) is a concept in physics that refers to the energy possessed by an object due to its position in a gravitational field.

The concept of gravitational potential energy is derived from the spatial arrangement of an object within a gravitational field, such as the one generated by the Earth. The determination of an object's potential energy relies on several factors, including the mass (m) of the object, the acceleration due to gravity (g), and its vertical displacement (h) from a chosen reference point, typically the ground. The equation representing gravitational potential energy is:

$$PE_G = mgh$$

Elastic Potential Energy:

Elastic potential energy is a form of energy that is linked to objects capable of undergoing deformation or compression, such as springs or elastic bands. When objects undergo stretching or compression, they accumulate elastic potential energy. The magnitude of elastic potential energy stored is contingent upon the object's spring constant (k) and the degree of deformation (x). The equation representing elastic potential energy is (Rajamäki et al., 2003):

$$PE_E = \frac{1}{2} kx^2$$

3.10.3. Conservation of Mechanical Energy

The fundamental concept of mechanical energy conservation asserts that the overall mechanical energy within a closed system stays constant without external forces except conservative forces such as gravity. Converting mechanical energy between different forms, such as kinetic and potential, is possible; however, the overall quantity of mechanical energy remains conserved (Pierratos & Polatoglou, 2017). Mathematically, the principle of conservation of mechanical energy can be formulated as follows:

$$KEi + PE\text{Gi} + PE\text{Ei} = Kef + PE\text{Gf} + PE_{Ef}$$

where:

KE is the system's kinetic energy, *PE*E is the elastic potential energy, and *PE*G is the gravitational potential energy. And *i* and *f* represent initial and final, respectively. Conservative forces, such as gravitation and elastic forces, exhibit path independence, indicating that the work performed by these forces solely relies on the starting point and ending point of the object rather than the specific trajectory followed during the displacement. The principle of conservation of mechanical energy remains valid for systems in which only conservative forces are present (Halilovic et al., 2022).

3.10.4. Power and Efficiency

The concepts of power and efficiency are crucial in understanding the rate at which energy is transferred and utilized in different systems.

3.10.4.1. Power

Power can be defined as the rate at which work is performed or when power is transferred or transformed. The concept pertains to the rate at which energy is consumed or generated. The watt (W) is the standard unit of power, defined as the equivalent of one joule per second (1 J/s). In mathematical terms, the concept of power (P) can be expressed as follows (Low et al., 2008):

$$P = \frac{W}{t}$$

P, W, and t represent the following quantities: P denotes power, W represents work done or energy transferred, and t signifies the time taken for the work or energy transfer. Power is a scalar quantity that conveys information regarding the rate at which energy is being consumed or supplied during a given process (He et al., 2012).

3.10.4.2. Efficiency

Efficiency measures how well a system converts input energy into useful output energy. It is the ratio of useful output energy (or work) to the total input energy (or work).

Mathematically, efficiency (η) is given by:

$$\eta = \frac{\text{Useful Output EnergyTotal Input Energy}}{\text{Total Input EnergyUseful Output Energy}} x100$$

Efficiency is commonly quantified in the form of a percentage. A system characterized by high efficiency is regarded as being more effective in converting input energy into useful work, thereby reducing energy loss in the form of waste.

Efficiency is a crucial factor in technology and engineering, where the primary goal in system and process design is to optimize efficiency. Minimizing energy losses resulting from friction, heat, or other forms of inefficiency can substantially enhance the overall operational efficiency of a given system (Heo et al., 2014).

3.11. MOMENTUM AND COLLISIONS

The concepts of momentum and collisions are basic concepts within classical mechanics that elucidate the dynamics of objects and their interactions. Momentum, a vector quantity, encompasses the magnitude and direction of an object's motion. Collisions, on the other hand, pertain to the exchange of momentum between multiple objects. This section will delve into the concepts of linear momentum and impulse, the principle of conservation of momentum, various classifications of collisions (elastic and inelastic), and practical applications of momentum in real-world scenarios (Ültay & Alev, 2017).

3.11.1. Linear Momentum and Impulse

a. **Momentum:**

Linear momentum, also called momentum, is a vector quantity that describes the motion of an object. The quantity in question is determined by the multiplication of the mass of an object (m) and its velocity (v), as expressed by the given equation:

Momentum (p)=$m \cdot v$

The unit of momentum in the International System of Units (SI) is expressed as kilogram-meter per second (kg m/s). Given that velocity is a vector quantity, it follows that momentum, being directly related to velocity, also possesses a direction that aligns with the direction of the object's velocity.

The concept of momentum pertains to the manifestation of an object's motion and serves as a quantifiable indicator of the resistance encountered when attempting to halt or alter said motion. The magnitude of an object's momentum directly correlates with the level of resistance encountered when attempting to modify its state of motion (Serevina et al., 2017).

b. **Impulse:**

The concept of impulse refers to the alteration in the momentum of an object, which occurs as a consequence of an external force being exerted upon it. The quantity in question is a vector, and its value can be determined using the following equation:

Impulse (J)=$\Delta p = pf - pi$

where: Δp is the change in momentum, pf is the final, and pi is the initial momentum. The nature of an impulse can vary, either positive or negative, contingent upon the direction in which the momentum change occurs. When a force is exerted in alignment with the initial velocity, the resulting impulse is positive, augmenting the object's momentum.

In contrast, when a force is exerted in the opposite direction of the initial velocity, the resulting impulse is negative, decreasing the object's momentum. The concept of impulse can be understood in terms of the force (F) exerted and the duration (Δt) during which the force is applied, as described by the impulse-momentum theorem:

Impulse (J)=$F \cdot \Delta t$

The theorem posits that the impulse encountered by an object is equivalent to the product of the force exerted on it and the duration of time during which the force is exerted (Hidayat & Dewantara, 2021).

3.11.2. Conservation of Momentum

The principle of momentum conservation is a fundamental concept in physics, positing that the total momentum of a closed system remains constant in the absence of external forces. In alternative terms, the principle of conservation of momentum posits that the total momentum before a collision or interaction is equal to the total momentum after the collision.

The principle of momentum conservation can be mathematically represented in the following manner:

According to the equation, the total initial momentum of all objects within a given system is equivalent to the total final momentum of all objects within the same system.

According to the equation, the total initial momentum of all objects within a given system is equivalent to the total final momentum of all objects within the same system (Caucao et al., 2020).

Within the confines of a closed system, the fundamental principle of momentum conservation asserts that the aggregate momentum of all entities engaged in a collision or interaction remains unaltered, whereby the initial momentum is equivalent to the ultimate momentum.

The principle above is derived from Newton's third law of motion, which states that for every action, there is an equal and opposite reaction. In collisions, the transfer of momentum between objects in interaction is exclusively attributed to internal forces without external forces acting upon the system.

The momentum conservation principle is highly effective for analyzing diverse physical situations, including collisions and explosions. This phenomenon enables the estimation of the ultimate velocities of entities engaged in collisions, contingent upon their initial velocities and masses (Jäger et al., 2003).

3.11.3. Types of Collisions

The categorization of collisions between objects can be broadly divided into two primary classifications: elastic collisions and inelastic collisions. The classifications above are derived from the principles governing the preservation of kinetic energy in the context of collisions (Petrović et al., 2020).

3.11.3.1. Elastic Collisions

In the context of an elastic collision, it is observed that both momentum and kinetic energy are preserved. Stated differently, the sum of the kinetic energy before the collision is equivalent to the sum after the collision (Thornton et al., 2011).

Elastic collisions are distinguished by the absence of any permanent deformation or energy dissipation throughout the interaction. When two objects experience an elastic collision, they exhibit a rebound or separation following the collision without any dissipation of mechanical energy. In the context of mathematical analysis, the principles of conservation of momentum and kinetic energy can be mathematically represented for an elastic collision between two objects denoted by subscripts 1 and 2 (Brunetto & Strazzulla, 2005):

Conservation of Momentum

$$m_1 \cdot v_{1i} + m_2 \cdot v_{2i} = m_1 \cdot v_{1f} + m_2 \cdot v_{2f}$$

Conservation of Kinetic Energy

$$\frac{1}{2} m_1 \cdot v_1^{i2} + \frac{1}{2} m_2 \cdot v_2^{i2} = \frac{1}{2} m_1 \cdot v_1^{f2} + \frac{1}{2} m_2 \cdot v_2^{f2}$$

3.11.3.2. Inelastic Collisions

In the context of an inelastic collision, the momentum conservation principle holds, ensuring that the system's total momentum remains constant. However, it should be noted that the conservation of kinetic energy is not fully upheld in such collisions. Mechanical energy can transform into various other forms, including internal energy, deformation, and heat.

Inelastic collisions are distinguished by the absence of complete separation or rebound between the objects involved. Following the collision, the objects can adhere to each other or exhibit collective motion (Stewart & Trinkle, 1996). In the context of mathematical analysis, the principle of momentum conservation in an inelastic collision between two objects, denoted by subscripts 1 and 2, can be mathematically represented as follows:

$m_1 \cdot v_{1i} + m_2 \cdot v_{2i} = (m_1 + m_2) \cdot v_{\text{final}}$ (Conservation of Momentum)

The final velocity (vfinal) of the combined objects in an inelastic collision can be determined by employing the conservation of momentum equation. Nevertheless, it should be noted that the principle of conservation of kinetic energy does not hold in this scenario, resulting in final kinetic energy that is lower than the initial kinetic energy. A particular instance of an inelastic collision is a perfectly inelastic collision, wherein the involved objects adhere to one another and travel with a shared final velocity. In this scenario, the total kinetic energy is maximally dissipated, resulting in the objects merging into a single consolidated mass after the collision (Fröbrich & Gontchar, 1998).

3.11.4. Applications of Momentum in Real-World Scenarios

The momentum principle and its preservation has extensive practical implications in various disciplines and industries. Presented below are several illustrative instances:

1. The comprehension of momentum and collisions plays a vital role in the development of car safety features. In a collision, the momentum conservation principle is employed to ascertain the forces at play and devise structures that can effectively absorb and dissipate energy, thereby mitigating the impact experienced by individuals involved in the collision (Osaba et al., 2021).
2. Sports: The concept of momentum holds considerable importance in various sports, including football, soccer, and rugby. In the context of a game, players strategically leverage their momentum to attain a competitive edge. The momentum conservation principle becomes apparent when players collide or alter their trajectory.
3. In explosions and projectile motion, momentum conservation is employed to examine the trajectory of fragments or debris generated by the detonation.
4. Astrophysics encompasses the study of celestial bodies, including planets, comets, and asteroids, and their motion and interactions. A fundamental concept in this field is the understanding of momentum and the conservation of momentum, which play a

crucial role in comprehending the dynamics of these celestial entities (Dimitriadis et al., 2016).

5. Spacecraft and Rocketry: The momentum conservation principle is employed in space missions to determine the necessary velocity and propellant mass adjustments for spacecraft to achieve desired orbits or destinations.
6. Particle colliders utilize the momentum conservation principle in particle physics to investigate the interactions and reactions occurring among subatomic particles (Bongs et al., 2019).

3.12. QUANTUM MECHANICS

Quantum mechanics, also known as quantum physics or quantum theory, pertains to the field of physics that investigates the characteristics of matter and energy at the microscopic level, where the principles of classical physics become inadequate. The theory in question emerged in the early 20th century as an attempt to elucidate the perplexing behaviors exhibited by subatomic particles. Over time, it has evolved into a highly successful and foundational framework within contemporary physics. In contrast to classical mechanics, which operates under deterministic principles and addresses continuous quantities, quantum mechanics incorporates probabilistic and discrete elements in its depiction of the natural world.

In this discourse, we shall delve into several fundamental principles of quantum mechanics that present a departure from classical intuitions. By doing so, we aim to enhance our comprehension of the underlying forces and dynamics operating at the quantum scale (Jammer, 1974).

3.12.1. Wave-Particle Duality

The concept of wave-particle duality is a notable and paradoxical feature within quantum mechanics. Particles such as electrons and photons were characterized in classical physics as discrete entities possessing well-defined positions and trajectories. In contrast, the classical wave theory elucidated light and other phenomena as continuous waves that spread through the medium of space.

Nevertheless, experiments carried out during the early 20th century, including the renowned double-slit experiment, demonstrated the ability of subatomic particles to manifest characteristics of particles and waves. The double-slit experiment involves the passage of a beam of electrons or photons through two closely spaced slits, observing an interference pattern on a screen. This pattern exhibits characteristics typically associated with waves.

The existence of this duality suggests that particles can demonstrate wave-like characteristics, while waves can also exhibit particle-like properties. This phenomenon challenges our conventional comprehension of the fundamental nature of matter and energy. The phenomenon of particles exhibiting wave-like behavior is elucidated through wave functions, which are mathematical constructs that offer insights into the likelihood of locating a particle within a specific state (Dimitrova & Weis, 2008).

3.12.2. Uncertainty Principle

The uncertainty principle, originally postulated by Werner Heisenberg, is a pivotal and foundational principle within the realm of quantum mechanics. The statement posits that specific pairs of physical properties exist, such as position and momentum, for which it is impossible to simultaneously ascertain both with complete precision, regardless of the level of accuracy sought.

Stated differently, a higher level of precision in determining one of these characteristics corresponds to a lower level of precision in determining the other.The mathematical formulation of the uncertainty principle is given by the inequality $\Delta x * \Delta p \geq \hbar/2$, where Δx represents the uncertainty in position, Δp represents the uncertainty in momentum, and $\hbar$ (h-bar) denotes the reduced Planck's constant, which is a fundamental constant in the field of physics.

The principle above possesses significant ramifications for the behavior exhibited by particles within the realm of quantum mechanics. This implies that particles lack clearly defined paths and exhibit intrinsically probabilistic behavior.

The presence of uncertainty gives rise to wave functions as probability distributions, which characterize the probability of locating a particle at a specific position with a specific momentum.

The uncertainty principle is not solely attributable to limitations in measurement technology but rather represents an inherent and fundamental characteristic of quantum particles. The phenomenon imposes a constraint on the precision of concurrent measurements of specific pairs of attributes (Ali et al., 2019).

3.12.3. Quantum Tunneling

The phenomenon of quantum tunneling is notable within the realm of quantum mechanics, wherein particles can traverse barriers traditionally considered impenetrable based on classical principles. According to the principles of classical physics, when a particle comes into contact with a barrier that possesses a height greater than the particle's energy, it is anticipated to undergo complete reflection. This is due to the particle's insufficient energy to overcome the barrier.

Did You Know?

Quantum tunneling allows particles to pass through barriers that, according to classical mechanics, would be insurmountable.

Nevertheless, particles exhibit distinct behavioral patterns within the domain of quantum mechanics. The phenomenon of quantum tunneling enables particles to traverse barriers, regardless of their energy being lower than the height of said barriers. This phenomenon arises as a result of the wave-like characteristics exhibited by particles. When a particle interacts with a barrier, its wave function exhibits wave-like behavior extending beyond it. This extension enables the particle to possess a non-zero probability of being detected on the opposite side of the barrier. Quantum tunneling exhibits a multitude of practical applications, including its utilization in the functioning of tunnel diodes, scanning tunneling microscopy (STM), and nuclear fusion. Furthermore, a comprehensive comprehension of phenomena occurring within stars, nuclear reactions, and the stability of atomic nuclei is of utmost importance.

The phenomenon of quantum tunneling is a captivating expression of the non-classical characteristics observed in the quantum realm, and it holds significant importance in determining the behavior of matter and energy at the atomic and subatomic levels.

The field of quantum mechanics has significantly transformed our comprehension of the underlying forces and dynamics operating at the minuscule levels of the natural world. The phenomenon under consideration presents a significant departure from conventional intuitions, unveiling the probabilistic, non-deterministic, and wave-like characteristics inherent in particles. The principles of wave-particle duality, the uncertainty principle, and quantum tunneling have revealed a complex and enigmatic quantum realm that diverges significantly from the familiar classical world encountered in our everyday experiences. The advancement of quantum mechanics has not only expanded our understanding of the microscopic realm. Still, it has also yielded various technological innovations, including quantum computing and cryptography, which can bring about transformative changes in numerous domains. Despite its remarkable achievements, Quantum mechanics engenders profound philosophical inquiries about the fundamental essence of reality, the influence of observers in measurement processes, and the inherent boundaries of human comprehension. As the investigation and expansion of quantum mechanics persist, it is anticipated that a more profound understanding of the principles and dynamics governing the quantum realm will be attained. This will consequently facilitate the exploration of novel domains in science and technology and the continued unraveling of enigmatic aspects within our universe (Gatteschi & Sessoli, 2003).

Activity 1.1

Consider a scenario where you are sliding a book on a table. Calculate the force of friction acting against the motion of the book given the mass of the book, the coefficient of kinetic friction, and the acceleration due to gravity. Also, discuss how this friction force influences the motion of the book.

SUMMARY

This chapter provides an in-depth exploration of the principles governing forces and motion. It begins by developing the concept of force, leading into Newton's Three Laws of Motion: the law of inertia, the law governing force and acceleration, and the law of action and reaction. The chapter further explores different types of forces such as normal force, tension, and gravitational force. It discusses further applications of Newton's laws and introduces the four basic forces. The concept of friction and drag forces, their types, and ways to reduce them are explored. The chapter also covers the principles of work, energy, power, and their interrelations, including the conservation of mechanical energy. Momentum and its conservation in collisions are explained, along with real-world applications. The chapter concludes with an introduction to quantum mechanics, discussing wave-particle duality, the Uncertainty Principle, and quantum tunneling.

REVIEW QUESTIONS

1. What is the concept of force and how did it develop historically?
2. Explain Newton's Three Laws of Motion. How do they govern the dynamics of objects?
3. What are the different units of force and how do they relate to the concept of weight and gravitational force?
4. Discuss the nature of normal force and tension. How do they operate in various physical scenarios?
5. How can friction and drag forces be reduced in real-world applications?
6. Explain the work-energy principle. How does it govern the relationship between work, energy, and power?
7. What is the principle of conservation of momentum? How does it apply in different types of collisions?
8. Discuss the principles of wave-particle duality and the Uncertainty Principle in quantum mechanics. What is quantum tunneling, and why is it significant?

REFERENCES

1. Abdulbari, H. A., Salleh, M. A. M., Rashed, M. K., & Ismail, M. H. S. (2018). Passive, active, and interactive drag-reduction technique to reduce friction and enhance the mixing intensity in rotating disk apparatus. *Chemical Engineering Communications*, *205*(12), 1623–1640.
2. Ali, A. F., Das, S., & Vagenas, E. C. (2009). Discreteness of space from the generalized uncertainty principle. *Physics Letters B*, *678*(5), 497–499.
3. Antwi, V. (2015). Using Real-Life Activities in an Interactive Engagement Manner in the Teaching and Learning of Newton's First Law of Motion in a Ghanaian University. *Journal of Education and Practice*, *6*(12), 48–58.
4. Bayraktar, Z., Komurcu, M., & Werner, D. H. (2010). Wind Driven Optimization (WDO): A novel nature-inspired optimization algorithm and its application to electromagnetics. In *2010 IEEE Antennas and Propagation Society International Symposium* (pp. 1–4). IEEE.
5. Birch, J. M., & Dickinson, M. H. (2003). The influence of wing–wake interactions on the production of aerodynamic forces in flapping flight. *Journal of Experimental Biology*, *206*(13), 2257–2272.
6. Bongs, K., Holynski, M., Vovrosh, J., Bouyer, P., Condon, G., Rasel, E., ... & Roura, A. (2019). Taking atom interferometric quantum sensors from the laboratory to real-world applications. *Nature Reviews Physics*, *1*(12), 731–739.
7. Brunetto, R., & Strazzulla, G. (2005). Elastic collisions in ion irradiation experiments: A mechanism for space weathering of silicates. *Icarus*, *179*(1), 265–273.
8. Bruun, J., & Brewe, E. (2013). Talking and learning physics: Predicting future grades from network measures and Force Concept Inventory pretest scores. *Physical Review Special Topics-Physics Education Research*, *9*(2), 020109.
9. Caucao, S., Oyarzúa, R., & Villa-Fuentes, S. (2020). A new mixed-FEM for steady-state natural convection models allowing conservation of momentum and thermal energy. *Calcolo*, *57*(4), 36.
10. Charnes, A., Clark, C. T., Cooper, W. W., & Golany, B. (1984). A developmental study of data envelopment analysis in measuring the efficiency of maintenance units in the US air forces. *Ann. Oper. Res.*, *2*(1), 95–112.

11. Chowdhury, M. A., & Helali, M. (2008). The effect of amplitude of vibration on the coefficient of friction for different materials. *Tribology International*, *41*(4), 307–314.

12. Cohen, I. B. (1964). 'Quantum in se est': Newton's Concept of Inertia in Relation to Descartes and Lucretius. *Notes and Records of the Royal Society of London*, *19*(2), 131–155.

13. Coletta, V. P., Phillips, J. A., & Steinert, J. J. (2007). Interpreting force concept inventory scores: Normalized gain and SAT scores. *Physical Review Special Topics-Physics Education Research*, *3*(1), 010106.

14. Czajko, J. (2022). Mathematical extension of special relativity into three-dimensional internal acceleration. *Physics Essays*, *35*(2), 165–170.

15. De Oliveira, R. R. L., Albuquerque, D. A. C., Cruz, T. G. S., Yamaji, F. M., & Leite, F. L. (2012). Measurement of the nanoscale roughness by atomic force microscopy: basic principles and applications. *Atomic Force Microscopy-Imaging, Measuring and Manipulating Surfaces at the Atomic Scale*, *3*.

16. de Pereira, A. P., Lima Junior, P., & Rodrigues, R. F. (2016). Explaining as mediated action: An analysis of pre-service teachers' account of forces of inertia in non-inertial frames of reference. *Science & Education*, *25*, 343–362.

17. Dimitriadis, P., Tegos, A., Oikonomou, A., Pagana, V., Koukouvinos, A., Mamassis, N., ... & Efstratiadis, A. (2016). Comparative evaluation of 1D and quasi-2D hydraulic models based on benchmark and real-world applications for uncertainty assessment in flood mapping. *Journal of Hydrology*, *534*, 478–492.

18. Dimitrova, T. L., & Weis, A. (2008). The wave-particle duality of light: A demonstration experiment. *American Journal of Physics*, *76*(2), 137–142.

19. Ding, J. (2023). The Theory on Thing's Limits. Part 2: A Brief Analysis of the New Knowledge of Newton's First Law. *Journal of Electronic & Information Systems*, *5*(1), 10–19.

20. Elo, M., Benjowsky, C., & Nummela, N. (2015). Intercultural competences and interaction schemes—Four forces regulating dyadic encounters in international business. *Industrial Marketing Management*, *48*, 38–49.

21. Epstein, J. (2007). Development and validation of the Calculus Concept Inventory. In *Proceedings of the Ninth International Conference on*

Mathematics Education in a Global Community (Vol. 9, pp. 165–170). Charlotte, NC.

22. Erfan, M., & Ratu, T. (2018). Analysis of student difficulties in understanding the concept of Newton's law of motion. *JIPF (Jurnal Ilmu Pendidikan Fisika)*, *3*(1), 1–4.
23. Fröbrich, P., & Gontchar, I. I. (1998). Langevin description of fusion, deep-inelastic collisions and heavy-ion-induced fission. *Physics Reports*, *292*(3–4), 131–237.
24. Gatteschi, D., & Sessoli, R. (2003). Quantum tunneling of magnetization and related phenomena in molecular materials. *Angewandte Chemie International Edition*, *42*(3), 268–297.
25. Gee, R., Hanley, C., Hussain, R., Canuel, L., & Martinez, J. (2015). Axial oscillation tools vs. Lateral vibration tools for friction reduction–what's the best way to shake the pipe?. In *SPE/IADC Drilling Conference and Exhibition* (p. D021S012R006). SPE.
26. Gibilaro, L. G., Di Felice, R., Waldram, S. P., & Foscolo, P. U. (1985). Generalized friction factor and drag coefficient correlations for fluid-particle interactions. *Chemical Engineering Science*, *40*(10), 1817–1823.
27. Goldie, P. A., Bach, T. M., & Evans, O. M. (1989). Force platform measures for evaluating postural control: reliability and validity. *Archives of Physical Medicine and Rehabilitation*, *70*(7), 510–517.
28. Halilovic, A., Mešić, V., Hasović, E., & Vidak, A. (2022). The Post-Instruction Conceptions about Conservation of Mechanical Energy: Findings from a Survey research with High School and University Students. *Journal of Turkish Science Education*, *19*(1), 144–162.
29. Hashiguchi, K., & Ozaki, S. (2008). Constitutive equation for friction with transition from static to kinetic friction and recovery of static friction. *International Journal of Plasticity*, *24*(11), 2102–2124.
30. Hashima, A., Yaakoba, O. B., Koha, K. K., Ismaila, N., & Ahmeda, Y. M. (2015). Review of Micro-bubble Ship Resistance Reduction Methods and the Mechanisms that Affect the Skin Friction on Drag Reduction from 1999 to 2015.
31. He, Z., Zhong, C., Su, S., Xu, M., Wu, H., & Cao, Y. (2012). Enhanced power-conversion efficiency in polymer solar cells using an inverted device structure. *Nature Photonics*, *6*(9), 591–595.

32. Hejazi, V., Sobolev, K., & Nosonovsky, M. (2013). From superhydrophobicity to icephobicity: forces and interaction analysis. *Scientific Reports*, *3*(1), 2194.

33. Hendricks, S. B. (1941). Base Exchange of the Clay Mineral Montmorillonite for Organic Cations and its Dependence upon Adsorption due to van der Waals Forces. *The Journal of Physical Chemistry*, *45*(1), 65–81.

34. Heo, J. H., Song, D. H., & Im, S. H. (2014). Planar CH3NH3PbBr3 hybrid solar cells with 10.4% power conversion efficiency, fabricated by controlled crystallization in the spin-coating process. *Advanced Materials*, *26*(48), 8179–8183.

35. Hestenes, D., & Halloun, I. (1995). Interpreting the force concept inventory: A response to March 1995 critique by Huffman and Heller. *The Physics Teacher*, *33*(8), 502–502.

36. Hidayat, R., & Dewantara, D. (2021). The development of electronic teaching materials on linear impulse and linear momentum to improve students' scientific literacy. In *Journal of Physics: Conference Series* (Vol. 2104, No. 1, p. 012011). IOP Publishing.

37. Hsu, V. W., Thayer, J. P., Wang, W., & Burns, A. (2016). New insights into the complex interplay between drag forces and its thermospheric consequences. *Journal of Geophysical Research: Space Physics*, *121*(10), 10–417.

38. Israelachvili, J. N., & McGuiggan, P. M. (1990). Adhesion and short-range forces between surfaces. Part I: New apparatus for surface force measurements. *Journal of Materials Research*, *5*, 2223–2231.

39. Jäger, N. D., Weber, E. R., Urban, K., & Ebert, P. (2003). Importance of carrier dynamics and conservation of momentum in atom-selective STM imaging and band gap determination of GaAs (110). *Physical Review B*, *67*(16), 165327.

40. Jammer, M. (1974). Philosophy of Quantum Mechanics. The interpretations of quantum mechanics in historical perspective, *2*(5), 1–8

41. Kotwal, C. A., & Bhushan, B. (1996). Contact analysis of non-Gaussian surfaces for minimum static and kinetic friction and wear. *Tribology Transactions*, *39*(4), 890–898.

42. Kulworawanichpong, T. (2015). Multi-train modeling and simulation integrated with traction power supply solver using simplified Newton–Raphson method. *Journal of Modern Transportation*, *23*, 241–251.

43. Legaye, J., & Duval-Beaupere, G. (2008). Gravitational forces and sagittal shape of the spine: clinical estimation of their relations. *International Orthopaedics*, *32*, 809–816.
44. Lemmer, M. (2018). Applying the science of learning to the learning of science: Newton's second law of motion. *Africa Education Review*, *15*(1), 20–37.
45. Liangzao, F., Jinsong, F., & Wuqing, L. (2010). An experiment discovery about gravitational force changes in materials due to temperature variation. *Engineer Sciences*, *8*(2), 9–11.
46. Liu, S. X., Li, Y. C., & Li, G. W. (2007). Wave current forces on the pile group of base foundation for the East Sea Bridge, China. *Journal of Hydrodynamics*, *19*(6), 661–670.
47. Low, Z. N., Chinga, R. A., Tseng, R., & Lin, J. (2008). Design and test of a high-power high-efficiency loosely coupled planar wireless power transfer system. *IEEE Transactions on Industrial Electronics*, *56*(5), 1801–1812.
48. Lubarda, M. V., & Lubarda, V. A. (2022). A review of the analysis of wind-influenced projectile motion in the presence of linear and nonlinear drag force. *Archive of Applied Mechanics*, *92*(7), 1997–2017.
49. Malm, J., Graff-Radford, N. R., Ishikawa, M., Kristensen, B., Leinonen, V., Mori, E., ... & Relkin, N. R. (2013). Influence of comorbidities in idiopathic normal pressure hydrocephalus—research and clinical care. A report of the ISHCSF task force on comorbidities in INPH. *Fluids and Barriers of the CNS*, *10*(1), 1–14.
50. Moo, J. G. S., Mayorga-Martinez, C. C., Wang, H., Teo, W. Z., Tan, B. H., Luong, T. D., ... & Pumera, M. (2018). Bjerknes forces in motion: Long-range translational motion and chiral directionality switching in bubble-propelled micromotors via an ultrasonic pathway. *Advanced Functional Materials*, *28*(25), 1702618.
51. Moody, C. A., Hebert, G. N., Strauss, S. H., & Field, J. A. (2003). Occurrence and persistence of perfluorooctanesulfonate and other perfluorinated surfactants in groundwater at a fire-training area at Wurtsmith Air Force Base, Michigan, USA. *Journal of Environmental Monitoring*, *5*(2), 341–345.
52. Morris, L., Delassus, P., Walsh, M., & McGloughlin, T. (2004). A mathematical model to predict the in vivo pulsatile drag forces acting on bifurcated stent grafts used in endovascular treatment of abdominal aortic aneurysms (AAA). *Journal of Biomechanics*, *37*(7), 1087–1095.

53. Nestler, M., & Voigt, A. (2021). Active nematodynamics on curved surfaces--the influence of geometric forces on motion patterns of topological defects. *arXiv preprint arXiv:2107.07779.*

54. Nieminen, P., Savinainen, A., & Viiri, J. (2010). Force Concept Inventory-based multiple-choice test for investigating students' representational consistency. *Physical Review Special Topics-Physics Education Research*, *6*(2), 020109.

55. Osaba, E., Villar-Rodriguez, E., Del Ser, J., Nebro, A. J., Molina, D., LaTorre, A., ... & Herrera, F. (2021). A tutorial on the design, experimentation and application of metaheuristic algorithms to real-world optimization problems. *Swarm and Evolutionary Computation*, *64*, 100888.

56. Papaxanthis, C., Pozzo, T., Popov, K. E., & McIntyre, J. (1998). Hand trajectories of vertical arm movements in one-G and zero-G environments Evidence for a central representation of gravitational force: Evidence for a central representation of gravitational force. *Experimental Brain Research*, *120*, 496–502.

57. Perrin, D. H., Robertson, R. J., & Ray, R. L. (1987). Bilateral isokinetic peak torque, torque acceleration energy, power, and work relationships in athletes and nonathletes. *Journal of Orthopaedic & Sports Physical Therapy*, *9*(5), 184–189.

58. Petrović, Đ., Mijailović, R., & Pešić, D. (2020). Traffic accidents with autonomous vehicles: type of collisions, manoeuvres and errors of conventional vehicles' drivers. *Transportation Research Procedia*, *45*, 161–168.

59. Pierratos, T., & Polatoglou, H. M. (2017). Study of the conservation of mechanical energy in the motion of a pendulum using a smartphone. *Physics Education*, *53*(1), 015021.

60. Planinic, M., Ivanjek, L., & Susac, A. (2010). Rasch model based analysis of the Force Concept Inventory. *Physical Review Special Topics-Physics Education Research*, *6*(1), 010103.

61. Podkletnov, E., & Nieminen, R. (1992). A possibility of gravitational force shielding by bulk YBa2Cu3O7− X superconductor. *Physica C: Superconductivity*, *203*(3–4), 441–444.

62. Pospíšil, J. (2013). Possible Uses of Newton's Laws of Motion in Commodity Price Theory and the Training of Expert Appraisers at Universities. *Procedia-Social and Behavioral Sciences*, *106*, 2071–2079.

63. Rajamäki, T., Miani, A., & Halonen, L. (2003). Vibrational energy levels for symmetric and asymmetric isotopomers of ammonia with an exact kinetic energy operator and new potential energy surfaces. *The Journal of Chemical Physics*, *118*(14), 6358–6369.
64. Saglam-Arslan, A., & Devecioglu, Y. (2010). Student teachers' levels of understanding and model of understanding about Newton's laws of motion. In *Asia-Pacific Forum on Science Learning & Teaching* (Vol. 11, No. 1).
65. Sanders, J. E., Greve, J. M., Mitchell, S. B., & Zachariah, S. G. (1998). Material properties of commonly-used interface materials and their static coefficients of friction with skin and socks. *Journal of Rehabilitation Research and Development*, *35*, 161–176.
66. Schwendeman, R. H. (1966). Comparison of experimentally derived and theoretically calculated derivatives of the energy, kinetic energy, and potential energy for CO. *The Journal of Chemical Physics*, *44*(5), 2115–2119.
67. Serevina, V., Nasbey, H., & Andriana, W. (2017). The Development of a learning material using the scratch programming language to helping student leaning momentum and impulse subject in senior high school. In *Proceedings of international Conference on Technology and Sosial Science*.
68. Shinogaya, T., Bakke, M., Thomsen, C. E., Vilmann, A., Sodeyama, A., & Matsumoto, M. (2001). Effects of ethnicity, gender and age on clenching force and load distribution. *Clinical Oral Investigations*, *5*, 63–68.
69. Sjökvist, L., Wu, J., Ransley, E., Engström, J., Eriksson, M., & Göteman, M. (2017). Numerical models for the motion and forces of point-absorbing wave energy converters in extreme waves. *Ocean Engineering*, *145*, 1–14.
70. Stewart, D. E., & Trinkle, J. C. (1996). An implicit time-stepping scheme for rigid body dynamics with inelastic collisions and coulomb friction. *International Journal for Numerical Methods in Engineering*, *39*(15), 2673–2691.
71. Sujarwanto, E., & Putra, I. A. (2018). Conception of motion as Newton law implementation among students of physics education. *Jurnal Pendidikan Sains*, *6*(4), 110–119.
72. Tawil, M., & Said, M. A. (2022). Understanding the Newton's Motion Concept Through Qualitative and Quantitative Teaching. *JPPPF (Jurnal Penelitian dan Pengembangan Pendidikan Fisika*, *8*(1).

73. Thompson, R. O. R. Y. (1980). Efficiency of conversion of kinetic energy to potential energy by a breaking internal gravity wave. *Journal of Geophysical Research: Oceans*, *85*(C11), 6631–6635.
74. Thornton, C., Cummins, S. J., & Cleary, P. W. (2011). An investigation of the comparative behaviour of alternative contact force models during elastic collisions. *Powder Technology*, *210*(3), 189–197.
75. Tsai, M. H. (2015). Application of the work-energy principle to assess the risetime effect on the dynamic response amplification under column loss *5*(2), 1–125
76. Ültay, E., & Alev, N. (2017). Investigating the Effect of the Activities Based on Explanation Assisted REACT Strategy on Learning Impulse, Momentum and Collisions Topics. *Journal of Education and Practice*, *8*(7), 174–186.
77. Wang, L., & Lu, K. (2023). Solving the general 3-D safety factor by combining Sarma's idea with the assumption of normal stress distribution over the slip surface. *Plos One*, *18*(6), e0287998.
78. Wayne, R. (2012). A Fundamental, Relativistic and Irreversible Law of Motion: A Unification of Newton's Second Law of Motion and the Second Law of Thermodynamics. *arXiv preprint arXiv:1206.3109*.
79. White, B. Y. (1984). Designing computer games to help physics students understand Newton's laws of motion. *Cognition and Instruction*, *1*(1), 69–108.
80. Wimmer, M., Regensburger, A., Bersch, C., Miri, M. A., Batz, S., Onishchukov, G., ... & Peschel, U. (2013). Optical diametric drive acceleration through action–reaction symmetry breaking. *Nature Physics*, *9*(12), 780–784.
81. Yifat, Y., Coursault, D., Peterson, C. W., Parker, J., Bao, Y., Gray, S. K., ... & Scherer, N. F. (2018). Reactive optical matter: light-induced motility in electrodynamically asymmetric nanoscale scatterers. *Light: Science & Applications*, *7*(1), 105.
82. Zammit, G. V., Menz, H. B., & Munteanu, S. E. (2010). Reliability of the TekScan MatScan® system for the measurement of plantar forces and pressures during barefoot level walking in healthy adults. *Journal of Foot and Ankle Research*, *3*, 1–9.
83. Zhang, Y., Zhang, G., Fierro, R., & Yang, Y. (2018). Force-driven traffic simulation for a future connected autonomous vehicle-enabled smart transportation system. *IEEE Transactions on Intelligent Transportation Systems*, *19*(7), 2221–2233.

CHAPTER 4

Work, Energy, and Power

Contents

4.1. INTRODUCTION

The concepts of work, energy, and power are fundamental in physics, serving as essential tools for comprehending the dynamics of physical systems and the various forms of energy conversion. Each concept possesses distinct characteristics and finds applications in diverse domains, making them indispensable constituents of classical mechanics and other scientific and engineering disciplines.

Work is the product of the distance an object covers with the force exerted on it. When an object undergoes work, energy is converted from one form to another. Work is a scalar quantity typically quantified in joules (J) within the International System of Units (SI) framework. The concept of work holds great significance due to its capacity to quantitatively measure the amount of effort applied to accomplish a particular displacement. The act of lifting an object, pushing a cart, or stretching a spring all involve the concept of work, which is an essential element in comprehending mechanical phenomena (Hubley & Wells, 1983).

Energy is an inherent characteristic of both objects and systems, serving as the basis for their ability to perform work. Energy can manifest in diverse forms, including but not limited to kinetic energy, potential energy, thermal energy, and other variations. According to the energy conservation principle, a closed system's total energy remains constant. Comprehending energy is paramount as it dictates the dynamics of particles, entities, and systems. The study of energy transformation and transfer enables us to understand how systems operate, encompassing various applications that impact various facets of human existence, including physiological processes and the functioning of intricate machinery (Moore, 2018).

Power can be defined as the rate at which work is performed or energy is transferred or transformed. The metric quantifies the rate at which energy is consumed or provided during a given process. Power is classified as a scalar quantity in physics, and its unit of measurement in the International System of Units (SI) is the watt (W). The watt is equivalent to one joule per second (1 J/s). The significance of power resides in its capacity to quantitatively measure the effectiveness and operational capabilities of systems and machines. Ahigh power output is highly sought in numerous applications, including engines and electrical devices, due to its indication of the capacity to execute work quickly and efficiently (Jacobson et al., 2017).

4.2. RELATIONSHIP BETWEEN WORK, ENERGY, AND POWER

Work, energy, and power are interrelated and describe distinct facets of energy and its various transformations.

The concept of work in physics refers to transferring energy from one form to another. An object's energy changes when work is performed on it. As an illustration, when an external force is exerted to elevate a book, work is performed, resulting in the transfer of potential energy to the book as it increases in vertical position. Likewise, in the case of an upwardly thrown ball, mechanical work is performed in opposition to the force of gravity, converting the ball's initial kinetic energy into potential energy when it reaches the apex of its trajectory. The work-energy principle is a fundamental concept that describes the connection between work and energy. It asserts that the amount of work exerted on an object is equivalent to the alteration in its kinetic energy. The principle above elucidates the inherent relationship between labor and energy conversion within a given system (Chen et al., 2011).

Power pertains to the rate at which work is performed, or energy is transferred. In mathematical terms, power can be defined as the rate of work change concerning time. The concept denotes the rate at which tasks are executed or the speed at which energy is transferred. The correlation between work and power can be comprehended using the theoretical construct of impulse. The concept of impulse refers to the alteration in the momentum of an object, which arises as a consequence of a force exerted on it during a defined period. The impulse-momentum theorem establishes a connection between impulse and work, stating that the impulse encountered by an object is equivalent to the work exerted on it (Pawlowski & Perrin, 1989).

4.3. WORK AND THE WORK-ENERGY PRINCIPLE

The concept of work in physics is a fundamental principle that describes the transfer of energy that occurs when a force is applied to an object, causing the object to undergo displacement in the same direction as the force (Figure 4.1). The concept quantifies the amount of work done to achieve a specific displacement, and it holds significant importance in mechanical processes and energy conversions.

The mathematical representation of work is defined as the dot product between the force vector (F) and the displacement vector (d):

Work (W)=F·d

In the given equation, the symbol W represents the amount of work performed, F denotes the magnitude of the force exerted on the object, and d signifies the object's displacement in the direction of the applied force.

It is imperative to acknowledge that work is a scalar quantity, signifying that it possesses magnitude but lacks direction. The joule (J) is the standard unit of work in the International System of Units (SI), and it is defined as the equivalent of one newton-meter (N·m) (Lian et al., 2021).

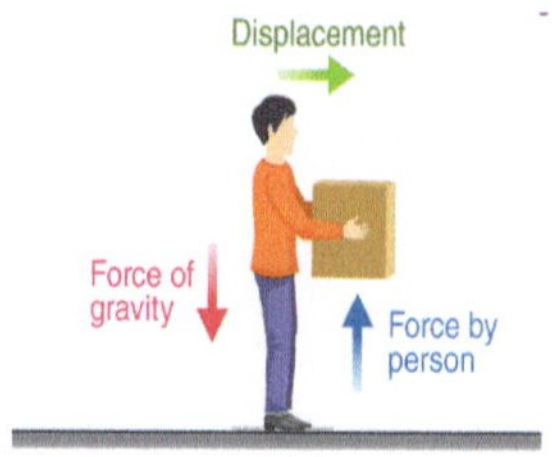

Figure 4.1. Illustration of work done (Source: Byju's Creative Commons License)

4.3.1. Work Done by Constant and Variable Forces

Determining work done by constant forces becomes straightforward when the force exerted on an object remains consistent over the entire displacement. In instances of this nature, the work performed can be mathematically represented as the multiplication of the constant force exerted on an object and the distance traversed by said object in the same direction as the force.

Tip

When calculating work done, consider whether the force applied is constant or variable as this will influence the method of calculation.

The equation W = F·d represents the relationship between work done, constant force applied to an object, and the object's displacement. In this equation, W denotes the work done, F represents the constant force, and d signifies the object's displacement.

For instance, when an individual applies a consistent force of 50 N to displace a box over a distance of 5 meters, the amount of work performed on the box can be determined by computing the product of the force and the displacement:

W=50 N·5 m=250J

The calculation of work done by variable forces involves dividing the displacement of an object into small intervals and multiplying the force acting on the object during each interval by the corresponding displacement. The cumulative work is determined by aggregating the work accomplished during each interval.

Mathematically, the work done by a variable force can be expressed as an integral:

$$W=\int\_d1{}^\wedge d2F\,dd$$

Where: W is the work done, F is the varying force acting on the object, d_1 and d_2 are the initial and final displacements, respectively (Czajko, 2014).

4.3.2. Positive and Negative Work

Determining the work's sign is crucial for comprehending the direction of energy transfer.

When the force exerted on an object aligns with the direction of displacement, resulting in an angle between them that is less than 90 degrees, the work performed is considered positive. This observation suggests an energy transfer to the object, increasing its kinetic energy (Figure 4.2) (Margaria, 1968).

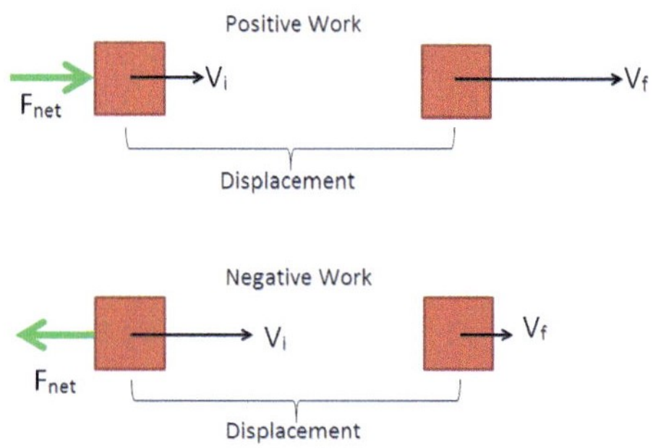

Figure 4.2. Illustration of positive and negative work (Source: Energy Education, Creative Commons License)

In contrast, when the force and displacement exhibit opposite directions (i.e., the angle between them exceeds 90 degrees), the work performed is characterized by a negative value. This suggests that energy is extracted from the object, resulting in a reduction in its kinetic energy. As an illustration,

1. When an individual exerts an upward force on a box, causing it to be lifted vertically, the force and displacement vectors align, leading to the generation of positive work and subsequently elevating the box's potential energy.
2. When an individual vertically lowers an object, the force exerted, and the object's displacement occurs in opposite directions, resulting in negative work and a reduction in the object's potential energy (Karatepe, 2010).

4.3.3. Work-Energy Principle and its Importance

The work-energy principle is a fundamental principle in mechanics that establishes a direct correlation between the amount of work exerted on an object and the resulting change in its kinetic energy. The concept of kinetic energy is derived from the fundamental principles of work and its definition.

The principle of work energy can be formulated in the following manner:

The amount of work performed on an object is equivalent to the alteration in its kinetic energy.

Mathematically, the work-energy principle can be expressed as follows:

$W=\Delta KE$

where: The variable W represents the work performed on the object, while ΔKE denotes the object's kinetic energy alteration.

The principle at hand underscores the inherent correlation between work and energy, underscoring that work serves to transfer energy to or from an object, thereby engendering a modification in its state of motion.

The significance of the work-energy principle resides in its extensive applicability for the analysis of mechanical systems and comprehension of energy conversions. The utilization of this tool proves to be highly effective in addressing issues about the dynamics of motion, collisions, and the interplay between various entities. Additionally, it offers a foundational framework for investigating energy conservation in diverse physical scenarios (Camarca et al., 2007).

4.3.4. Examples and Applications of the Work-Energy Principle

Applying the work-energy principle in simple systems can be illustrated by examining the scenario of a ball being projected vertically upward. As the ball ascends, it encounters the force of gravity exerted in a direction opposite to its displacement. The gravitational force performs negative work on the ball, signifying a reduction in its kinetic energy as it undergoes deceleration caused by gravity. It momentarily ceases its motion at the zenith of the ball's trajectory. At this juncture, the gravitational work is rendered null due to the perpendicular alignment of displacement and force, with an angle of 90 degrees (Pride et al., 1998).

Applying the work-energy principle can be observed in oscillatory motion, such as a simple pendulum. As the pendulum's mass undergoes periodic swinging, the tension in the string, which serves as the exerted force, performs work on the mass. During the nadir of the swing, the tension force and displacement exhibit parallel alignment, yielding positive work and augmenting the pendulum's kinetic energy. During the peak points of the swing, there exists a scenario where the tension force and displacement exhibit opposing directions, resulting in negative work. At these specific points, the pendulum momentarily ceases its motion and completely converts its kinetic energy into potential energy. The work-energy principle elucidates the ongoing exchange of kinetic and potential energy during the oscillatory motion of a pendulum.

The work-energy principle is a fundamental concept in the study of mechanical systems. The work-energy principle is widely utilized in the analysis of mechanical systems involving forces, such as springs and inclined planes. When a spring is compressed or stretched, work is performed on the spring, leading to the accumulation of elastic potential energy. Upon the release of the spring, the potential energy is subsequently transformed into kinetic energy as the spring undergoes oscillation. In the scenario involving the descent of an object on an inclined plane, the gravitational force exerts work on the object while it traverses the inclined surface. The gravitational force performs mechanical work on the object, converting its potential energy, which is determined by its vertical position relative to the ground, into kinetic energy as it accelerates while sliding (Lian et al., 2022).

The application of the work-energy principle in the context of machines and engines. The comprehension of the functioning of machines and engines is fundamentally rooted in the work-energy principle. In mechanical systems, such as internal combustion engines or steam turbines, combustion generates high-pressure gases. These gases, in turn, apply a force onto pistons or blades, facilitating the desired mechanical output. The exertion of this force leads to work performance, facilitating the conversion of thermal energy derived from the fuel into mechanical energy, thereby propelling the machinery. Furthermore, the conversion of mechanical energy into electrical energy occurs in electric generators through the utilization of magnetic fields, which are generated by the rotational motion of turbines. Within the realm of motors, a distinct phenomenon occurs known as the reverse process, wherein electrical energy is converted into mechanical work (Mustofa et al., 2019).

4.4. KINETIC ENERGY AND POTENTIAL ENERGY

4.4.1. Kinetic Energy

The concept of kinetic energy is a fundamental principle in the field of physics, which quantifies the amount of energy an object possesses as a result of its motion. Kinetic energy is one of the two primary forms of mechanical energy alongside potential energy (Figure 4.3). The concept of kinetic energy is closely linked to an object's mass and velocity, and it serves as a fundamental factor in comprehending the dynamics of moving entities:

$$KE = \frac{1}{2} mv^2$$

where: *KE* is the kinetic energy, *m* is the mass, and *v* is the object's velocity.

Figure 4.3. Examples of kinetic energy (Source: Engineerine, Creative Commons License)

Kinetic energy is classified as a scalar quantity, indicating that it possesses magnitude but lacks direction. The joule (J) is the standard international (SI) unit used to measure kinetic energy. It is defined as the product of the mass of an object and the square of its velocity, expressed in units of kilogram meter squared per second squared (kg m^2/s^2) (Schwendeman, 1966).

4.4.2. Potential Energy

Potential energy is a variant manifestation of mechanical energy that signifies the energy that an object possesses due to its specific position or arrangement within a force field. Various forms of potential energy exist, among which gravitational and elastic potential energy are frequently encountered.

The concept of gravitational potential energy refers to the energy possessed by an object due to its position in a gravitational field:

$PE = mgh$

where: PE is the gravitational potential energy, m is the mass, g is the gravitational acceleration, and h is the height.

The relationship between gravitational potential energy and the mass of an object, as well as its height relative to a reference level, is one of direct proportionality (Figure 4.4). The potential energy of an object increases as it is raised to a higher position (Seino et al., 2019).

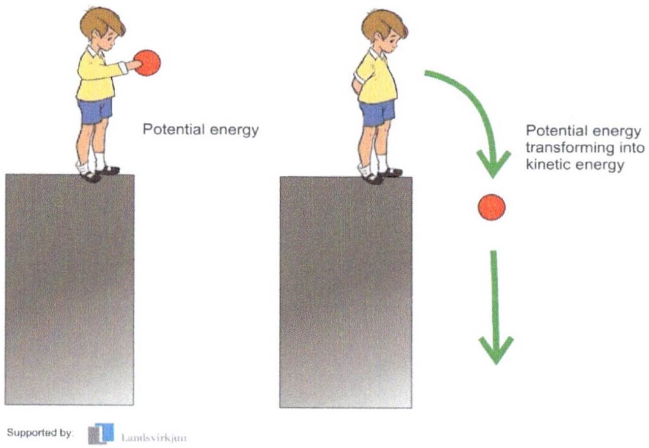

Figure 4.4. Illustration of the potential energy (Source: Polar Pedia, Creative Commons License)

The concept of elastic potential energy refers to the energy stored in an elastic material when deformed from its original shape. Elastic potential energy is a form of energy that is attributed to objects capable of undergoing deformation or elongation, such as springs or elastic bands. When an external force is exerted on an object, causing it to undergo compression or elongation, the object accumulates potential energy. The elastic potential energy of an object increases in direct proportion to the degree of compression or stretching it undergoes.

The equation representing the concept of elastic potential energy is expressed as:

Elastic Potential Energy (PE)=$1/2\,kx^2$

where: *PE* is the elastic potential energy; *k* is k is the force constant of spring; *x* is the displacement of the object from its mean position (Rajamäki et al., 2003).

4.4.3. Relationship between Kinetic and Potential Energy

The comprehension of the interaction between kinetic energy and potential energy is of utmost importance in elucidating the dynamics of mechanical energy within a given system. The principle of conservation of mechanical energy posits that the overall mechanical energy of an enclosed system remains unaltered when solely conservative forces, such as gravity and elastic forces, are exerted upon the system.

E=KE+PE

This principle is demonstrated in a variety of situations. As an illustration, a ball being projected upwards acquires potential energy as it ascends owing to its augmented vertical displacement from the ground.

At the apex of its trajectory, the object's kinetic energy reaches a minimum value, while its potential energy attains its maximum value. When the ball undergoes descent, the potential energy is converted into kinetic energy, while the overall mechanical energy remains constant, assuming there is no presence of air resistance (Allen et al., 2007).

4.4.4. Conversion between Kinetic and Potential Energy

The interconversion of kinetic and potential energy is fundamental to numerous physical phenomena. The mechanical energy of an object undergoing motion within a force field undergoes a continuous transformation between kinetic energy and potential energy, contingent upon the object's position and velocity. The conversion from potential to kinetic energy occurs when an object is situated at a greater elevation, resulting in increased potential energy. As an object descends under the force of gravity, its potential energy diminishes while its kinetic energy augments. The phenomenon of conversion can be observed in situations such as releasing an object from an elevated position or the descent of a roller coaster down a slope.

The conversion of potential energy into kinetic energy adheres to the principle of conservation of mechanical energy. As the potential energy of a system decreases, an equal amount of kinetic energy is acquired, resulting in the overall mechanical energy of the system remaining constant (Liu et al., 2015).

The process of converting kinetic energy into potential energy. In contrast, when an object undergoes upward motion within a force field, such as vertically throwing a ball, there is a decrease in its kinetic energy while its potential energy experiences an increase. At the apex of its trajectory, the object exhibits a complete absence of kinetic energy, with its entire mechanical energy being stored in the form of potential energy (Chau & Mukherjee, 2019).

4.4.5. Real-World Examples of Kinetic and Potential Energy

The pendulum is a compelling illustration of the dynamic relationship between kinetic and potential energy. The pendulum exhibits periodic alterations in its position and velocity as it oscillates back and forth. At the nadir of its oscillation, the pendulum exhibits maximal kinetic and minimal potential energy. At the apices of its oscillation, the pendulum experiences a temporary cessation of motion, resulting in the dissipation of its kinetic energy to zero. At these specific locations, the entirety of its energy exists in potential energy, reaching its utmost level. When the pendulum descends, potential energy is converted into kinetic energy. A bouncing ball can be understood as a manifestation of energy conversion between kinetic and potential forms. During the ball's descent, potential energy is converted to kinetic energy. When an object undergoes a bounce, there is a momentary decrease in its kinetic energy to zero, and all of its energy is subsequently converted into potential energy. The conversion above process persists as the ball oscillates in a vertical motion (Figure 4.5) (Stephanik & Shaffer, 2012).

Springs play a significant role in spring-mass systems, wherein a mass is connected to a spring and displaced from its state of equilibrium, accumulating elastic potential energy.

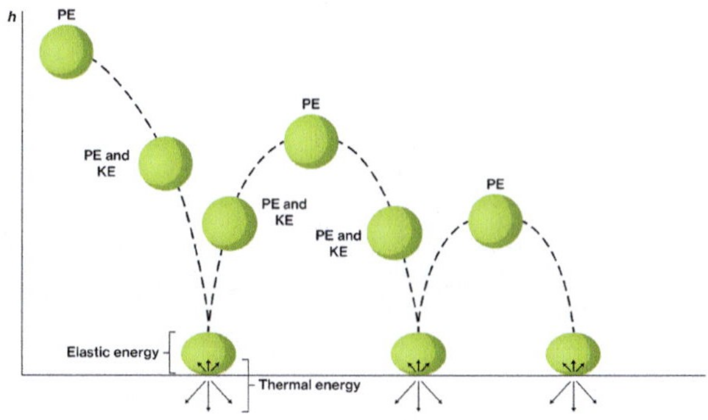

Figure 4.5. A ball released from a height, h, will convert potential energy to kinetic energy as it falls toward Earth. At the point of impact, most of the kinetic energy is converted to elastic energy, which causes the ball to bounce back up, but some kinetic energy is converted to thermal energy (Source: Course Hero, Creative Commons License).

During the oscillatory motion, the mass converts potential energy to kinetic energy when it reaches the midpoint. The conservation of mechanical energy is observed throughout the entirety of the oscillation. Hydroelectric power plants harness the potential energy of water that is stored at an elevated position and convert it into kinetic energy through turbines. The turbines facilitate the conversion of the kinetic energy inherent in the motion of water into mechanical energy, subsequently harnessed to generate electrical energy (Sun et al., 2018).

4.5. CONSERVATION OF MECHANICAL ENERGY

The principle of conservation of mechanical energy is a fundamental concept in physics, which asserts that the total mechanical energy of a closed system remains constant when subjected solely to conservative forces. But, a system's total energy, comprising both kinetic and potential energy, remains constant during its various motions or interactions as long as there are no external non-conservative forces, such as friction or air resistance (Karapetkov et al., 2019).

Remember

An isolated system where only conservative forces act, the total mechanical energy (kinetic + potential) remains constant.

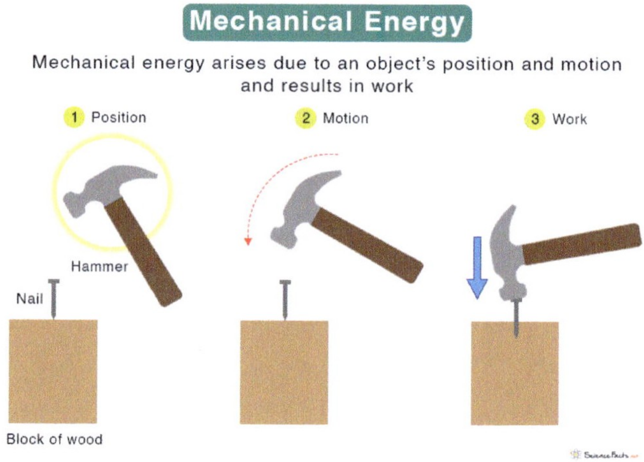

Figure 4.6. Schematic of the mechanical energy (Source: Science Facts, Creative Commons Licensse).

Mathematically:

Total Mechanical Energy E=Kinetic Energy (KE)+Potential Energy (PE)

4.5.1. Applications of Conservation of Mechanical Energy

The principle of conserving mechanical energy finds extensive applications across diverse domains within physics and engineering. Several notable applications of this technology include:

Applying the principle of conservation of mechanical energy greatly facilitates the comprehension of object motion within conservative force fields. In the case of a pendulum, the interconversion of kinetic energy and potential energy adheres to the principle of conservation of mechanical energy, resulting in the pendulum's total mechanical energy remaining constant throughout its oscillations. This principle is also applicable to the phenomenon of simple harmonic motion and the behavior of objects subjected to gravitational or elastic forces (Halilovic et al., 2022).

The mechanical energy conservation principle is paramount in the design and functioning of roller coasters and various other forms of amusement rides. Engineers employ this principle to guarantee that the roller coaster cars possess adequate kinetic energy to successfully traverse the ride while simultaneously considering potential energy variations at various locations along the track. The primary objective is to provide riders with a thrilling experience while prioritizing their safety.

The study of satellite orbits and other celestial bodies in celestial mechanics involves the application of the principle of conservation of mechanical energy. As an illustration, a satellite in orbit around a celestial body undergoes both gravitational potential energy and kinetic energy as a result of its orbital motion. Energy conservation is a fundamental principle for the entirety of a satellite's orbit. This principle plays a crucial role in predicting and analyzing orbital motion (Puttharugsa & Pimanpang, 2021).

The principle of mechanical energy conservation is employed in diverse renewable energy technologies. Hydropower plants facilitate the conversion of potential energy possessed by water, which is stored at an elevated position, into kinetic energy using its descent through turbines (Luis Díaz et al., 2008).

4.5.2. Examples of Conservation of Mechanical Energy in Simple Systems

Falling Object: Let us contemplate a rudimentary scenario involving a spherical object released from an elevated position. As the ball descends under the force of gravity, its potential energy diminishes while its kinetic energy augments. At the moment of release, the ball exhibits its maximum potential energy while its kinetic energy is at a minimum value of zero. As the ball descends toward the ground, there is a decrease in its potential energy, accompanied by a corresponding increase in kinetic energy. Upon reaching the ground, the ball's potential energy is completely dissipated, resulting in the complete conversion of energy into kinetic form.

Throughout the motion, the ball's aggregate mechanical energy, encompassing both potential and kinetic energy, remains constant due to the absence of non-conservative forces, such as air resistance, exerting any influence upon it (Halilovic et al., 2021).

The spring-mass system is a commonly studied physical system in mechanics. It consists of a mass connected to a spring. In the context of a mass-spring system, introducing a mass connected to a spring and subsequently displaced from its state of equilibrium results in the acquisition of potential energy. During the oscillatory motion, the mass converts potential energy to kinetic energy when it reaches the midpoint. At the limits of its oscillatory motion, where the displacement reaches zero, the potential energy attains its maximum value, while the kinetic energy diminishes to zero. When the mass returns to the midpoint, the potential energy is reduced to zero, resulting in the complete conversion of energy into kinetic energy.

The system's overall mechanical energy remains conserved throughout the oscillation, provided non-conservative forces such as friction are absent (Wangdi et al., 2017).

4.5.3. Non-Conservative Forces and Energy Dissipation

The principle of conservation of mechanical energy remains valid solely in scenarios where non-conservative forces, which can dissipate energy from the system, are absent. Non-conservative forces, including but not limited to friction, air resistance, and drag, result in the dissipation of mechanical energy within the system, ultimately transforming it into heat or other forms of energy. When a vehicle is in motion, the engine supplies the requisite mechanical energy to counteract the forces of friction and air resistance, enabling the vehicle to sustain its velocity. As the vehicle undergoes motion, a portion of its mechanical energy is dissipated as thermal energy. This dissipation occurs due to the friction between the tires and the road surface and the resistance encountered by the surrounding air against the vehicle's movement. As a result, the overall mechanical energy of the vehicle diminishes, leading to a deceleration unless the engine sustains the provision of supplementary energy (Lu et al., 2022). In situations involving the presence of non-conservative forces, the principle of conservation of mechanical energy is not applicable, resulting in the alteration of the system's energy over a given period. The energy above dissipation underscores the significance of considering non-conservative forces in practical scenarios and engineering implementations (Novik & Coveney, 1998).

4.6. WORK-ENERGY THEOREM AND NON-CONSERVATIVE FORCES

The work-energy theorem is a fundamental principle in physics that establishes a direct relationship between the work done on an object and the change in its kinetic energy. It is a powerful tool for analyzing the motion of objects subjected to external forces and is instrumental in understanding energy transformations in various physical scenarios.

Mathematically, work (W) is given by:

$$W=F\cdot d$$

The work-energy theorem is derived from Newton's second law of motion:

$F_{net}=m \cdot a$

Considering that acceleration is the change in velocity (Δv) per unit time (Δt), we can express it as:

$a=\Delta v/\Delta t$

Substituting this expression for acceleration into Newton's second law, we get:

$F_{net}=m \cdot \Delta v/\Delta t$

Now, we can rewrite the change in velocity (Δv) as the final velocity (vf) minus the initial velocity (vi):

$\Delta v=v_f-v_i$

Substituting this back into the equation for net force, we obtain:

$F_{net}=m \cdot v_f-v_i/\Delta t$

Next, we can rewrite the change in time (Δt) as the final time (t_f) minus the initial time (t_i):

$\Delta t=t_f-t_i$

Substituting this into the equation for net force, we get:

$F_{net}=m \cdot\ v_f-v_i/t_f-t_i$

Now, the average velocity (v^-) during the time interval t_f-t_i is given by:

$v^-=v_f+v_i/2$

Substituting this into the equation for net force, we have:

" $F_{net}=m \cdot\ v_f-v_i/\ t_f-t_i = m \cdot v-v_i/\ t_f-t_i$"

The work done on the object (W) can be expressed as the force (F_{net}) multiplied by the displacement (d):

$W=F_{net} \cdot d$

Substituting the expression for net force, we get:

$W=m \cdot\ v^- -v_i/\ t_f-t_i \cdot d$

The average velocity (v^-) is the total displacement (Δd) divided by the change in time (Δt):

$v^-=\Delta d/\Delta t$

Substituting this into the equation for work, we have:

$W=m \cdot\ \Delta t \Delta d-v_i/t_f-t_i \cdot d$

Now, the change in displacement (Δd) is the final displacement (df) minus the initial displacement (di):

$\Delta d = df - di$

Substituting this into the equation for work, we get:

$$W = m \cdot \frac{\frac{d_f - d_i}{\Delta t} - v_i}{t_f - t_i} \cdot d$$

The term $\frac{d_f - d_i}{\Delta t}$ represents the average velocity ($\bar{v}$) during the time interval $t_f - t_i$. Substituting this into the equation for work, we obtain:

$$W = m \cdot \frac{\bar{v} - v_i}{t_f - t_i} \cdot d$$

Finally, we recognize that the term $\frac{\bar{v} - v_i}{t_f - t_i}$ is the average acceleration ($\bar{a}$) during the time interval $t_f - t_i$. Substituting this into the equation for work, we arrive at the work-energy theorem:

$$W = m \cdot \bar{a} \cdot d$$

The work-energy theorem states that the work done on an object is equal to the product of its mass, the average acceleration during the time interval, and the object's displacement. This theorem directly links work done and changes in the object's kinetic energy (Yang, 2019).

Applications of the Work-Energy Theorem: The work-energy theorem finds numerous applications in various physical scenarios, including:

1. Projectile Motion: The work-energy theorem is used to analyze the motion of projectiles, such as balls thrown into the air or objects launched from cannons. It helps determine the projectile's velocity, height, and range.
2. Collisions: In collisions between objects, the work-energy theorem is used to understand the energy transfer between objects and the resulting changes in their velocities.
3. Simple Machines: The principle is employed to analyze the efficiency and mechanical advantage of simple machines, such as pulleys, levers, and ramps.

4. Energy Conservation: The work-energy theorem is essential in studying energy conservation in mechanical systems. It helps understand how mechanical energy is transformed and conserved in various physical processes (Figure 4.7) (Mahmudah & Sulistyaningsih, 2022).

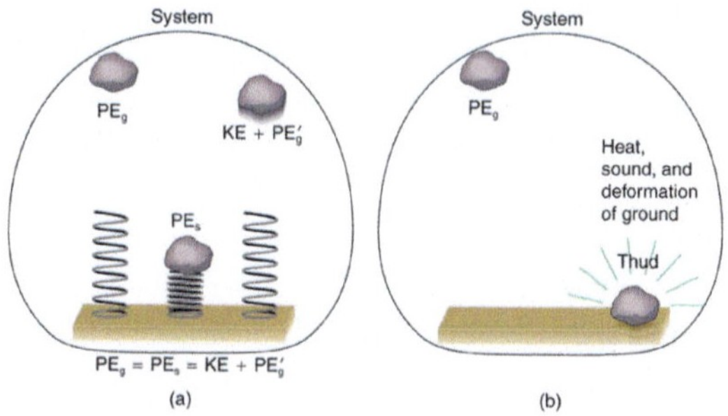

Figure 4.7. Comparison of the effects of conservative and nonconservative forces on the mechanical energy of a system (Source: Lumen Learning, Creative Commons License)

4.6.1. Friction as a Non-conservative Force

Friction is a prevalent illustration of a non-conservative force. The force opposes the object's motion's direction and facilitates mechanical energy's dissipation, resulting in its conversion into heat and sound. In contrast to conservative forces, which can retain and transfer energy within a system without experiencing any loss, non-conservative forces are responsible for energy dissipation from the system (Loi et al., 2011).

4.6.2. Other Non-conservative Forces and Their Effects

Apart from friction, several other forces are considered non-conservative due to their energy-dissipating nature. Some common examples include:

1. Air Resistance: When an object moves through the air, it experiences air resistance, which opposes its motion. Air resistance dissipates mechanical energy from the system, causing a decrease in the object's kinetic energy (Lazopoulos & Lazopoulos, 2010).
2. Drag Force: In fluid mechanics, objects moving through a fluid experience a drag force, which also opposes their motion. The work done by the drag force results in energy dissipation from the system.

3. Damping Forces: In oscillatory systems, damping forces, such as those in a damped harmonic oscillator, dissipate energy from the system, leading to a decrease in amplitude over time.
4. Viscous Forces: In fluid flow, viscous forces cause energy dissipation due to friction between fluid layers, leading to a loss of mechanical energy in the system (Wang & Rakich, 2011).

4.6.3. Role of Work-Energy Theorem in Solving Complex Problems

The work-energy theorem is a versatile tool for solving complex problems involving the motion of objects under the influence of various conservative and non-conservative forces. Its applications include:

1. Energy Analysis in Systems: The work-energy theorem helps analyze the energy flow in mechanical systems, including converting potential energy to kinetic energy and vice versa. It provides a comprehensive understanding of the energy changes in the system.
2. Efficiency of Machines: Engineers use the work-energy theorem to study the efficiency of machines and mechanical systems. They can optimize designs and reduce energy losses by considering the work done and energy dissipation.
3. Predicting Motion: In scenarios where forces are acting on an object, the work-energy theorem allows the prediction of the object's motion, velocity, and displacement.
4. Impact Analysis: During collisions, the work-energy theorem aids in understanding the transfer of kinetic energy and the outcome of the collision.
5. Design and Engineering: In engineering applications, the work-energy theorem is valuable for calculating forces and energy in structures, analyzing mechanical systems, and ensuring safety in design (Takahashi, 2019).

4.7. APPLICATIONS IN ENGINEERING AND TECHNOLOGY

The principles of work, energy, and power play a vital role in engineering and technology, especially in the design and analysis of mechanical systems.

Mechanical engineers use these principles to understand and optimize the performance of various machines and devices. Some key applications include:

1. Machine Design: Work and energy considerations are crucial in the design of machines to ensure efficiency and safety. Engineers calculate the work done and energy transfers within the machine to determine factors like required power, component sizing, and energy losses.
2. Efficiency Analysis: Engineers assess the efficiency of mechanical systems by evaluating the work input and output, as well as the energy losses due to friction, heat, and other factors. Improving efficiency helps minimize energy waste and operating costs.
3. Gear Systems: Work and energy principles are fundamental in designing gear systems, such as those found in gearboxes and transmissions. Engineers analyze these systems' power transmission, energy losses, and efficiency to optimize their performance.
4. Braking Systems: Understanding work and energy is crucial for designing effective braking systems in vehicles and machinery. Engineers consider the energy dissipation during braking and ensure that the system can efficiently handle the energy conversion and dissipation (Ali, 2011).

4.7.1. Renewable Energy and Energy Efficiency

With a growing focus on sustainability and environmental concerns, the applications of work, energy, and power in renewable energy technologies and energy efficiency have become increasingly important. Some notable applications include (Ponce & Khan, 2021):

1. Solar Energy: Work and energy principles are involved in converting sunlight into electricity through photovoltaic cells. Engineers analyze the work done by photons on the cell's semiconductor material to understand energy generation and efficiency (Leal Filho et al., 2019).
2. Wind Energy: The work-energy theorem plays a role in wind turbine design, where the kinetic energy of moving air is harnessed to generate electricity. Engineers analyze the work done on the turbine blades by the wind to optimize the turbine's efficiency.
3. Hydropower: In hydropower plants, the work-energy theorem is used to analyze falling water's potential and kinetic energy,

which is then converted into mechanical energy and, eventually, electrical energy (Naimoğlu & Akal, 2023).

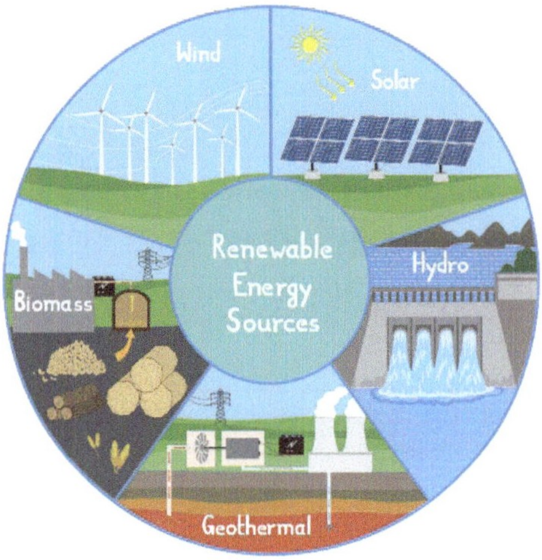

Figure 4.8. Types of renewable energies (Source: Shikha, Creative Commons License)

4. Energy Storage: Understanding work and energy principles is essential in designing efficient energy storage systems, such as batteries and flywheels. Engineers analyze energy input, storage capacity, and efficiency to improve energy storage technologies (Figure 4.8) (Lo, 2014).

4.7.2. Work and Power in Transportation and Vehicles

The principles of work and power are integral in transportation and vehicle engineering, influencing vehicle design, performance, and fuel efficiency. Some key applications include:

1. Internal Combustion Engines: In gasoline and diesel engines, work and energy principles are used to analyze the combustion process, energy conversion, and power output. Engineers optimize engine design to improve fuel efficiency and reduce emissions.
2. Electric Vehicles: Work and energy considerations are fundamental in design and battery management. Engineers analyze the work done by electric motors, energy storage, and power distribution systems to enhance vehicle range and efficiency (Clairand et al., 2019).

3. Hybrid Vehicles: A combination of internal combustion engines and electric motors is used in hybrid vehicles. Engineers analyze the work done and energy transfer between the two systems to maximize fuel efficiency and reduce environmental impact (Gboney, 2009).
4. Vehicle Dynamics: Work and energy principles play a role in understanding vehicle dynamics, such as braking, acceleration, and turning. Engineers use these principles to optimize vehicle stability and handling.

Did You Know?

Harnessing renewable energy sources such as wind and solar is a key strategy in increasing energy efficiency and reducing dependence on fossil fuels?

4.7.3. Power Generation and Distribution Systems

Work, energy, and power principles are essential for efficient and reliable electricity supply in power generation and distribution systems. Some key applications include:

1. Thermal Power Plants: In fossil fuel and nuclear power plants, work and energy principles generate steam to drive turbines and produce electricity (Qiu et al., 2022).
2. Electrical Grids: Work and power are fundamental in the transmission and distribution of electricity through power grids. Engineers analyze power losses, line efficiency, and load management to ensure stable and efficient power delivery (Alizadeh et al., 2016).
3. Renewable Power Integration: The work-energy theorem is used to understand energy generation and conversion in renewable power plants like solar, wind, and hydropower. Engineers analyze power output and variability to effectively integrate renewable energy sources into the grid (Guo et al., 2014).
4. Smart Grids: In modern smart grids, work and energy principles are utilized to optimize energy usage, manage demand, and improve grid efficiency. Engineers use advanced monitoring and control systems to balance power generation and consumption (Lojowska et al., 2011).

Activity 1.1

Imagine you're riding a roller coaster. At the highest point of the ride, you have maximum potential energy and minimum kinetic energy. As you descend, the potential energy gets converted into kinetic energy. Given the mass of the roller coaster and the height of the highest point, calculate the speed of the roller coaster at the bottom of the ride, assuming no energy losses due to friction or air resistance.

SUMMARY

This chapter delves into the fundamental concepts and relationships that govern the movement and behavior of objects. The chapter begins with an introduction to the relationship between work, energy, and power, followed by a detailed discussion of the work-energy principle. The concepts of positive and negative work are explained, along with real-world applications of the work-energy principle. Kinetic and potential energy are then explored, including their relationship and conversion between each other. The chapter emphasizes the conservation of mechanical energy and provides examples of this principle in various systems. Non-conservative forces like friction are examined, with a focus on the work-energy theorem's role in complex problems. The final section illustrates applications of these concepts in engineering and technology, including renewable energy, transportation, and power distribution systems.

REVIEW QUESTIONS

1. What is the relationship between work, energy, and power? How are these concepts applied in physics?
2. Explain the work-energy principle. What are positive and negative work, and how do they influence the work done on an object?
3. Describe the nature of kinetic and potential energy. How do they relate to each other, and what are some real-world examples of their conversion?
4. Discuss the conservation of mechanical energy. How do non-conservative forces like friction affect energy conservation in a system?
5. Explain the work-energy theorem and its application to non-conservative forces. How does it help in solving complex problems?

6. What are the applications of work, energy, and power in engineering and technology? Provide examples from renewable energy sources and transportation systems.
7. Describe the principles of power generation and distribution systems. How do these concepts impact energy efficiency?
8. Discuss the role of renewable energy in increasing energy efficiency. How does the utilization of renewable energy sources such as wind and solar influence the overall energy landscape?

REFERENCES

1. Ali, M. (2011). Coconut fiber: A versatile material and its applications in engineering. *Journal of Civil Engineering and Construction Technology*, *2*(9), 189–197.
2. Alizadeh, M., Wai, H. T., Chowdhury, M., Goldsmith, A., Scaglione, A., & Javidi, T. (2016). Optimal pricing to manage electric vehicles in coupled power and transportation networks. *IEEE Transactions on Control of Network Systems*, *4*(4), 863–875.
3. Allen, A. P., Gillooly, J. F., & Brown, J. H. (2007). Recasting the species-energy hypothesis: the different roles of kinetic and potential energy in regulating biodiversity. *Scaling Biodiversity*, *1*.
4. Camarca, M., Bonanno, A., & Sapia, P. (2007). Revisiting work–energy theorem's implications. *European Journal of Physics*, *28*(6), 1181.
5. Chau, S., & Mukherjee, R. (2019). Kinetic to potential energy transformation using a spring as an intermediary: application to the pole vault problem. *Journal of Applied Mechanics*, *86*(5), 051001.
6. Chen, G., Ployhart, R. E., Thomas, H. C., Anderson, N., & Bliese, P. D. (2011). The power of momentum: A new model of dynamic relationships between job satisfaction change and turnover intentions. *Academy of Management Journal*, *54*(1), 159–181..
7. Clairand, J. M., Guerra-Terán, P., Serrano-Guerrero, X., González-Rodríguez, M., & Escrivá-Escrivá, G. (2019). Electric vehicles for public transportation in power systems: A review of methodologies. *Energies*, *12*(16), 3114.
8. Czajko, J. (2014). Operationally complete work done suggests presence of extra potentials corresponding to repulsive forces. *International Letters of Chemistry, Physics and Astronomy*, *18*.
9. Gboney, W. (2009). Policy and regulatory framework for renewable energy and energy efficiency development in Ghana. *Climate Policy*, *9*(5), 508–516.
10. Guo, Q., Xin, S., Sun, H., Li, Z., & Zhang, B. (2014). Rapid-charging navigation of electric vehicles based on real-time power systems and traffic data. *IEEE Transactions on Smart Grid*, *5*(4), 1969–1979.
11. Halilovic, A., Mešic, V., Hasovic, E., & Dervic, D. (2021). Students' Difficulties in Applying the Law of Conservation of Mechanical Energy: Results of a Survey Research. *European Educational Researcher*, *4*(2), 171–192.

12. Halilovic, A., Mešić, V., Hasović, E., & Vidak, A. (2022). The Post-Instruction Conceptions about Conservation of Mechanical Energy: Findings from a Survey research with High School and University Students. *Journal of Turkish Science Education*, *19*(1), 144–162.
13. Hubley, C. L., & Wells, R. P. (1983). A work-energy approach to determine individual joint contributions to vertical jump performance. *European Journal of Applied Physiology and Occupational Physiology*, *50*, 247–254.
14. Jacobson, M. Z., Delucchi, M. A., Bauer, Z. A., Goodman, S. C., Chapman, W. E., Cameron, M. A., ... & Yachanin, A. S. (2017). 100% clean and renewable wind, water, and sunlight all-sector energy roadmaps for 139 countries of the world. *Joule*, *1*(1), 108–121
15. Karapetkov, S., Dimitrov, L., Uzunov, H., & Dechkova, S. (2019). Identifying vehicle and collision impact by applying the principle of conservation of mechanical energy. *Transport and Telecommunication Journal*, *20*(3), 191–204.
16. Karatepe, O. M. (2010). The effect of positive and negative work-family interaction on exhaustion: does work social support make a difference?. *International Journal of Contemporary Hospitality Management*, *22*(6), 836–856.
17. Lazopoulos, K. A., & Lazopoulos, A. (2010). Stability of a gradient elastic beam compressed by non-conservative forces. *ZAMM-Journal of Applied Mathematics and Mechanics/Zeitschrift für Angewandte Mathematik und Mechanik: Applied Mathematics and Mechanics*, *90*(3), 174–184.
18. Leal Filho, W., Salvia, A. L., Do Paco, A., Anholon, R., Quelhas, O. L. G., Rampasso, I. S., ... & Brandli, L. L. (2019). A comparative study of approaches towards energy efficiency and renewable energy use at higher education institutions. *Journal of Cleaner Production*, *237*, 117728.
19. Lian, R. Z., Guo, X. Y., & Xia, M. Y. (2021). Work-energy principle based characteristic mode theory with solution domain compression for metal-material composite scattering systems. *IEEE Transactions on Antennas and Propagation*, *70*(5), 3628–3642.
20. Lian, R. Z., Guo, X. Y., & Xia, M. Y. (2022). Entire-structure-oriented work-energy theorem (ES-WET)-based characteristic mode theory for material scattering objects. *IEEE Transactions on Antennas and Propagation*, *70*(7), 5699–5714.

21. Liu, G., Leng, Q., Lian, J., Guo, H., Yi, X., & Hu, C. (2015). Notepad-like triboelectric generator for efficiently harvesting low-velocity motion energy by interconversion between kinetic energy and elastic potential energy. *ACS Applied Materials & Interfaces*, *7*(2), 1275–1283.
22. Lo, K. (2014). A critical review of China's rapidly developing renewable energy and energy efficiency policies. *Renewable and Sustainable Energy Reviews*, *29*, 508–516.
23. Loi, D., Mossa, S., & Cugliandolo, L. F. (2011). Non-conservative forces and effective temperatures in active polymers. *Soft Matter*, *7*(21), 10193–10209.
24. Lojowska, A., Kurowicka, D., Papaefthymiou, G., & Van Der Sluis, L. (2011). From transportation patterns to power demand: Stochastic modeling of uncontrolled domestic charging of electric vehicles. In *2011 IEEE Power and Energy Society General Meeting* (pp. 1–7). IEEE.
25. Lu, D., Song, Z., Wang, G., Zhou, X., & Du, X. (2022). A developed 3D peridynamic method incorporating non-conservative force for brittle materials. *Engineering Fracture Mechanics*, *274*, 108772.
26. Luis Díaz, J., Avendaño, R. J., & Cuesta, Y. J. (2008). DIDACTIC APPLICATION OF A MECHANICAL DEVICE FOR TEACHING CONSERVATION OF MECHANICAL ENERGY, IN BASIC EDUCATION STUDENTS.
27. Mahmudah, I., & Sulistyaningsih, D. (2022). The Development of Laboratory-Scaled Mini Roller Coaster Media for Non-Conservative Energy Observations. In *Proceedings of the 3rd International Conference of Science Education in Industrial Revolution 4.0, ICONSEIR 2021, December 21st, 2021, Medan, North Sumatra, Indonesia*.
28. Margaria, R. (1968). Positive and negative work performances and their efficiencies in human locomotion. *Internationale Zeitschrift für angewandte Physiologie Einschließlich Arbeitsphysiologie*, *25*(4), 339–351.
29. Moore, J. W. (2018). The Capitalocene Part II: accumulation by appropriation and the centrality of unpaid work/energy. *The Journal of Peasant Studies*, *45*(2), 237–279.

30. Mustofa, Z., Sutopo, S., Mufti, N., & Asmichatin, A. (2019). The impact of modeling instruction based on system toward work-energy concept understanding. *Jurnal Penelitian & Pengembangan Pendidikan Fisika*, *5*(2), 145–154.
31. Naimoğlu, M., & Akal, M. (2023). The relationship between energy technology, energy efficiency, renewable energy, and the environment in türkiye. *Journal of Cleaner Production*, 138144.
32. Novik, K. E., & Coveney, P. V. (1998). Finite-difference methods for simulation models incorporating non-conservative forces. *arXiv preprint cond-mat/9808130*.
33. Pawlowski, D., & Perrin, D. H. (1989). Relationship between shoulder and elbow isokinetic peak torque, torque acceleration energy, average power, and total work and throwing velocity in intercollegiate pitchers. *Athletic Training*, *24*(2), 129–132.
34. Ponce, P., & Khan, S. A. R. (2021). A causal link between renewable energy, energy efficiency, property rights, and CO2 emissions in developed countries: a road map for environmental sustainability. *Environmental Science and Pollution Research*, *28*, 37804–37817.
35. Pride, T. O. B., Vokos, S., & McDermott, L. C. (1998). The challenge of matching learning assessments to teaching goals: An example from the work-energy and impulse-momentum theorems. *American Journal of Physics*, *66*(2), 147–157.
36. Puttharugsa, C., & Pimanpang, S. (2021). Analyzing the rotational motion of a rectangular board via smartphone sensors: a conservation-of-mechanical-energy approach. *Physics Education*, *56*(2), 023005.
37. Qiu, D., Wang, Y., Sun, M., & Strbac, G. (2022). Multi-service provision for electric vehicles in power-transportation networks towards a low-carbon transition: A hierarchical and hybrid multi-agent reinforcement learning approach. *Applied Energy*, *313*, 118790.
38. Rajamäki, T., Miani, A., & Halonen, L. (2003). Vibrational energy levels for symmetric and asymmetric isotopomers of ammonia with an exact kinetic energy operator and new potential energy surfaces. *The Journal of Chemical Physics*, *118*(14), 6358–6369.
39. Schwendeman, R. H. (1966). Comparison of experimentally derived and theoretically calculated derivatives of the energy, kinetic energy, and potential energy for CO. *The Journal of Chemical Physics*, *44*(5), 2115–2119.

40. Seino, J., Kageyama, R., Fujinami, M., Ikabata, Y., & Nakai, H. (2019). Semi-local machine-learned kinetic energy density functional demonstrating smooth potential energy curves. *Chemical Physics Letters*, *734*, 136732.
41. Stephanik, B. M., & Shaffer, P. S. (2012). Examining student ability to interpret and use potential energy diagrams for classical systems. In *AIP Conference Proceedings* (Vol. 1413, No. 1, pp. 367–370). American Institute of Physics.
42. Sun, H., Zhang, Y., Baleanu, D., Chen, W., & Chen, Y. (2018). A new collection of real world applications of fractional calculus in science and engineering. *Communications in Nonlinear Science and Numerical Simulation*, *64*, 213–231.
43. Takahashi, T. T. (2019). A Bad Moon Rising: The Puzzling Inaccuracies of the Work-Energy Theorem in Aircraft Performance. In *AIAA Scitech 2019 Forum* (p. 1305).
44. Wang, Z., & Rakich, P. (2011). Response theory of optical forces in two-port photonics systems: a simplified framework for examining conservative and non-conservative forces. *Optics Express*, *19*(22), 22322–22336.
45. Wangdi, D., Kanthang, P., & Precharattana, M. (2017). Development of a hands-on model embedded with guided inquiry laboratory to enhance students' understanding of law of mechanical energy conservation. In *Asia-Pacific Forum on Science Learning and Teaching* (Vol. 18, No. 2, pp. 1–27). Hong Kong Institute of Education. 10 Lo Ping Road, Tai Po, New Territories, Hong Kong.
46. Yang, Y. (2019). Study on Application of New Kinetic Energy Theorem of Rigid Body Fixed-Axis Rotation. In *IOP Conference Series: Materials Science and Engineering* (Vol. 563, No. 3, p. 032048). IOP Publishing.

CHAPTER 5

Momentum and Impulse and Collision

Contents

5.1. INTRODUCTION

The concept of momentum is a fundamental principle in physics, which describes the measure of motion inherent in an object. The quantity under consideration possesses the characteristics of a vector, thereby indicating that it possesses both magnitude and direction. The concept of momentum (p) in physics is characterized by the multiplication of an object's mass (m) and its velocity (v). In mathematical terms, the concept of momentum is represented by an expression (Han, 2015):

$$p=m\cdot v$$

where: p = momentum of an object, m = mass, and v = velocity.

The unit of momentum is commonly expressed as kilogram-meter per second (kg m/s), which measures the quantity of motion inherent in an object. An object possessing a substantial mass and exhibiting a considerable velocity will possess a momentum of greater magnitude in comparison to an object characterized by a smaller mass or a lower velocity (Triyani et al., 2019).

5.2. IMPULSE AND ITS RELATIONSHIP WITH MOMENTUM

The concept of impulse is also of great significance to momentum. The term "it" refers to the concept of change in momentum that occurs when an object is subjected to the influence of an external force. The vector quantity of impulse (J) can be mathematically represented as the product of force (F) and the duration of time (t) during which the force is exerted:

$$J=F\cdot t$$

The symbol "J" represents impulse, which is measured in units of Newton-seconds (Ns) or kilogram meters per second (kg m/s). The symbol F represents the force exerted on the object, measured in newtons (N), while the symbol t denotes the time the force is applied, measured in seconds (s). The principle of the impulse-momentum theorem posits that the magnitude of the impulse exerted on an object is equivalent to the alteration in its momentum. In a mathematical context, it can be expressed as (Pride et al., 1998):

$$J=\Delta p$$

where: Δp = change in momentum.

The theorem above postulates that applying a net force on an object over a specific period will result in the object undergoing an impulse, thereby causing a modification in its momentum. The principle of conservation of momentum is demonstrated by the constant preservation of momentum within a system when no external forces are acting upon it (Schilling et al., 2008).

5.3. LINEAR MOMENTUM

The concept of linear momentum is a fundamental principle in physics, which characterizes an object's motion along a straight trajectory. The term "momentum" refers to a vector quantity that is determined by multiplying the mass (m) of an object by its velocity (v).

The kilogram-meter per second (kg m/s) is the accepted unit for measuring linear momentum, which denotes the magnitude of an object's motion in a specific direction. Momentum, being a vector quantity, possesses both magnitude and direction. In the context of motion along a linear path, specifically in one dimension, the direction indication is determined by the polarity of the momentum. Positive momentum signifies the presence of movement in a singular direction, whereas negative momentum signifies movement in the opposite direction. The principle of conservation of momentum states that in the absence of external forces acting on a system, the linear momentum of the system remains constant. The conservation of momentum principle asserts that the overall momentum of a closed system remains constant throughout time, implying that the aggregate momentum of all entities within the system remains unchanged unless external forces are exerted (Belinfante, 1940).

5.3.1. Conservation of Linear Momentum

The principle of conservation of linear momentum is a fundamental concept in physics that finds extensive utility in various scenarios, encompassing collisions and interactions involving objects. Based on this principle, within a closed system devoid of external forces, the aggregate linear momentum before an occurrence is equivalent to the aggregate linear momentum after the occurrence. The conservation of linear momentum can be mathematically expressed as follows:

The principle of conservation of momentum states that a system's total initial momentum is equal to that system's total final momentum. The principle of conservation of momentum states that the total initial momentum of a system is equal to the total final momentum of the system (Hnizdo, 1992). The principle above is derived from Newton's third law of motion, which posits that a corresponding reaction exists for each action that is equal in magnitude but opposite in direction. During a collision, the forces exerted between two objects that collide are considered internal forces within the closed system. The internal forces acting on the objects produce equal and opposite impulses, modifying their respective momenta. Nevertheless, the overall momentum remains unchanging because the cumulative impulse within the enclosed system is equal to zero (Figure 5.1).

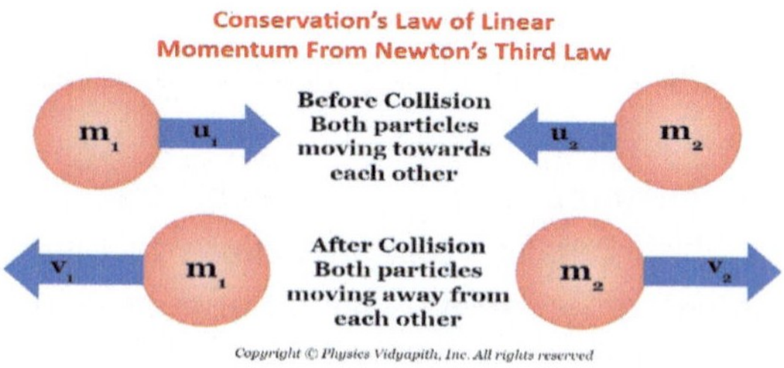

Figure 5.1. Conservation's law of linear momentum from Newton's Third Law (Source: Vidyapith, Creative Commons License)

The conservation of linear momentum principle is a valuable tool for analyzing and forecasting the results of different collision scenarios, including elastic and inelastic collisions (Scalora et al., 2006).

5.3.2. Elastic and Inelastic Collisions

The categorization of collisions between objects can be determined by the preservation or lack thereof of kinetic energy throughout the collision, resulting in the classification of elastic or inelastic collisions (Figure 5.2).

In the context of collisions, elastic collisions are characterized by conserving both momentum and kinetic energy. When two objects undergo an elastic collision, they exhibit a rebounding behavior where there is no dissipation of kinetic energy. The sum of the kinetic energies before the collision is equivalent to the sum after the collision. To illustrate, let us examine the scenario of two billiard balls colliding upon a surface devoid of friction. During an elastic collision, the initial ball imparts its momentum to the second ball, resulting in the second ball acquiring a velocity equivalent to that of the first ball before the collision. Both balls maintain their initial kinetic energies following the collision (Xie & Liu, 2020).

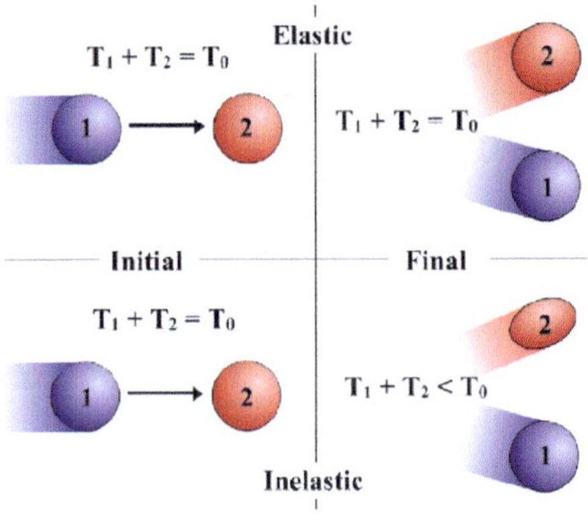

Figure 5.2. Comparison between elastic and inelastic collision (Source: Ian R Kleckner, Creative Commons License)

Tip

In elastic collisions, kinetic energy is conserved, while in inelastic collisions, kinetic energy is not conserved but momentum is still conserved.

In the context of collisions, specifically inelastic collisions, it is observed that while momentum is conserved, kinetic energy is not conserved. A portion of the kinetic energy undergoes conversion into alternative forms of energy, including but not limited to thermal energy, acoustic energy, or structural deformation of the objects involved in the collision. For example, let us consider a scenario in which two automobiles experience a collision. During an inelastic collision, it is possible for the vehicles involved to adhere to one another after the collision, leading to a reduction in the overall kinetic energy due to deformation and friction. Despite the principle of conservation of momentum, a portion of the energy is dissipated into the surrounding environment as non-mechanical energy (Engelhardt & Phelps, 1963).

5.3.3. Applications of Linear Momentum

Projectile motion is a well-known illustration of the principles of linear momentum. When an object is projected into the atmosphere and solely influenced by gravitational force, its horizontal velocity remains consistent due to the lack of external forces acting upon it. Nevertheless, the vertical velocity of the object changes, resulting in its movement along a parabolic path. The momentum conservation principle enables the comprehension and anticipation of projectile motion in diverse contexts, including sports, ballistics, and space exploration (Kuhl & Ramm, 1996).

In the realm of car crashes and safety engineering, the preservation of linear momentum holds significant importance. During a collision, the magnitude of momentum alteration endured by the individuals inside the vehicle can be significant.

The principle discussed herein is employed by engineers in the design of safety features, such as crumple zones, seat belts, and airbags, to mitigate the impact forces experienced by passengers, thereby minimizing the occurrence of injuries. Sports and athletics heavily rely on linear momentum, which holds significant importance.

Athletes use the momentum principle to attain high velocities, execute impressive leaps, and achieve remarkable projectile distances. As an illustration, sprinters strategically accumulate momentum throughout a race to attain maximum velocity. In contrast, long and high jumpers harness momentum to propel themselves through the air (Dehong et al., 1994).

The conservation of momentum plays a crucial role in rocket propulsion within the context of space exploration. Rockets operate based on Newton's third law, which states that the ejection of exhaust gases in a specific direction creates an equivalent and opposing force, thereby propelling the rocket in the opposite direction. The conservation of momentum enables rockets to attain substantial velocities and overcome Earth's gravitational pull, facilitating their journey into outer space (Meyers & Meneveau, 2013).

5.4. IMPULSE

The concept of impulse is a fundamental principle in physics, which quantifies the momentum alteration that an object undergoes when an external force acts upon it. The quantity in question is a vector, and its mathematical representation involves the multiplication of force (F) by the duration (t) during which the force is applied:

Impulse $(J)=F\cdot t$

The unit of impulse is commonly expressed as the Newton-second (Ns) or kilogram-meter per second (kg m/s), which is considered equivalent to the unit of linear momentum. The concept of impulse describes the alteration in momentum experienced by an object. It can be defined as the product of the force exerted on the object and the duration over which the force is applied.

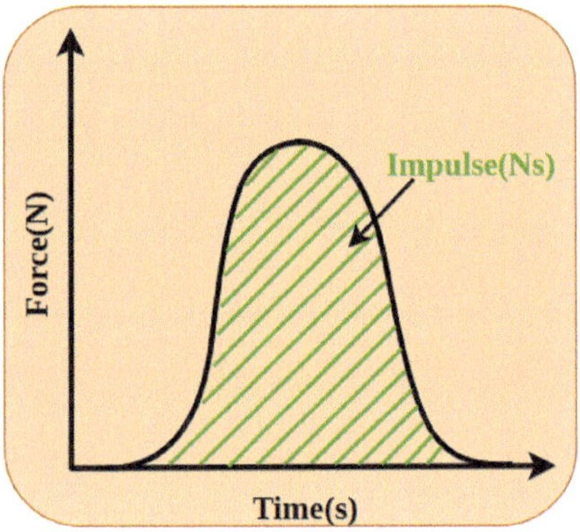

Figure 5.3. Schematic representation of the impulse (Source: Anurag, Creative Commons License)

A collision or an interaction involving a force results in the object experiencing an impulse, which subsequently causes a modification in its momentum. The magnitude of the impulse and the resulting change in momentum is directly proportional to the duration of time during which a force is applied (Figure 5.3) (Anders, 2011).

5.4.1. Relationship between Impulse and Change in Momentum

The correlation between impulse and change in momentum can be deduced from Newton's second law of motion, which postulates that the net force (Fnet) exerted on an object is equivalent to the product of its mass (m) and its acceleration (a):

$F_{net}=m \cdot a$

The acceleration of an object is:

$a = \Delta v/\Delta t.$

Combining:

$F_{net}=m \cdot \Delta v/\Delta t.$

Rearranging:

$F_{net} \cdot \Delta t=m \cdot \Delta v$

The quantity obtained by multiplying force and time, Fnet·Δt, is called impulse (J), while the alteration in velocity, Δv, is synonymous with the alteration in momentum (Δp). Hence, it is possible to formulate:

$$J = \Delta p$$

This relationship demonstrates that the impulse an object encounters is equivalent to the alteration in its momentum. When an object is exposed to a net force over a specific period, it will experience a modification in its momentum that is directly proportional to its impulse (Schilling et al., 2008).

5.4.2. Impulse-Momentum Theorem

The impulse-momentum theorem is a fundamental principle in physics that formalizes the inherent connection between impulse and the resulting change in momentum. According to the principle of impulse momentum, the magnitude of the impulse exerted on an object is equivalent to the alteration in its momentum. Mathematically, the impulse-momentum theorem can be formulated as follows (Kim et al., 2019):

$$J = \Delta p$$

5.4.3. Impulse and Force in Collision Analysis

In collision analysis, the concept of impulse plays a crucial role in comprehending the impact of forces involved in the collision phenomenon. Each object's momentum alteration in a collision is contingent upon the impulse exerted by the other object. The impulse-momentum theorem facilitates establishing a connection between the force exerted during a collision and the subsequent alteration in momentum.

In the context of elastic collisions, wherein both momentum and kinetic energy are conserved, the impulse encountered by each object can be determined by multiplying the force of interaction by the duration of the collision. Given that the total kinetic energy remains constant, it can be inferred that there is no energy dissipation within the system, and the forces at play are internal. By examining the impulses encountered by the colliding entities, scholars can ascertain the velocities and orientations of said entities subsequent to the collision. In the context of collisions, specifically inelastic collisions, it is observed that the conservation of kinetic energy does not hold. Instead, some initial energy is converted into different forms, such as thermal energy or deformation. Each object's momentum change is still influenced by the impulse experienced. However, the decrease in kinetic energy is accounted for by analyzing the external forces exerted on the objects throughout the collision (Fukui et al., 2021).

5.4.4. Safety Considerations and Design Implications

The concept of impulse holds considerable importance when considering safety factors in diverse engineering and real-world contexts. A comprehensive comprehension of the correlation between impulse and the alteration in momentum is paramount in developing safety mechanisms and preventive measures to mitigate the magnitude of impact forces endured by entities or individuals involved in collisions.

The severity of impact forces on occupants in car crashes is determined by the impulse experienced during a collision. To optimize passenger safety, engineers incorporate crumple zones into vehicle designs.

These zones are specifically engineered to undergo deformation in the event of a collision, effectively prolonging the duration of the collision and consequently mitigating the force experienced by the vehicle occupants. Furthermore, seat belts and airbags are specifically engineered to prolong the collision duration and evenly disperse the impact force, mitigating the likelihood of sustaining injuries (Zohar, 1980).

Remember

When designing safety features like airbags in cars, engineers consider the impulse experienced during collisions to reduce the impact forces on occupants.

Sports and Athletics: Within sports, a comprehensive comprehension of impulse plays a pivotal role in enabling coaches and athletes to enhance their performance to its maximum potential while concurrently mitigating the likelihood of sustaining injuries.

For example, athletes participating in sports such as gymnastics or diving strive to prolong their duration of contact with the ground or diving board upon landing to mitigate the impact forces exerted on their bodies. Similarly, safety equipment, such as helmets and padding, has been specifically engineered to effectively absorb and prolong the impact duration, thereby mitigating the impulse encountered in collisions or falls.

Incorporating impulse considerations is of paramount importance in engineering design, particularly in developing structures and materials that possess high resistance to impact. In aerospace engineering, it is imperative to consider the impact of impulse and forces encountered during various dynamic maneuvers, such as takeoff and landing, when designing aircraft components. In addition, the process of developing protective gear for industrial workers entails the optimization of impulses encountered during potential incidents or collisions. Safety standards and regulations frequently integrate impulse considerations to guarantee the safety of various products and structures. Engineers and regulators employ mathematical models and simulations to evaluate the impulse encountered in different scenarios and validate adherence to safety standards (Duthinh & Simiu, 2010).

5.5. COLLISIONS AND CONSERVATION OF MOMENTUM

5.5.1. Types of Collisions

Collisions are occurrences in which two or more entities interact, leading to their velocities alterations. Collisions can be classified into distinct categories according to the momentum and kinetic energy conservation principles exhibited during the interaction. Three primary categories of collisions exist elastic collisions, inelastic collisions, and perfectly inelastic collisions.

In the context of collisions, elastic collisions are characterized by conserving both momentum and kinetic energy. The aggregate kinetic energy of the system before the collision is equivalent to the collective kinetic energy after the collision. In an elastic collision, the objects involved undergo a rebound or separation following the interaction, without any energy dissipation into non-mechanical manifestations such as heat or deformation (McGuire & Kouri, 1974).

In an ideal elastic collision, the conditions that are satisfied are as follows:

1. The objects do not undergo deformation or encounter any energy dissipation resulting from internal friction.
2. The collision transpires without any external forces exerted on the system.
3. Elastic collisions are frequently observed in the interactions involving atomic and subatomic particles, as exemplified in particle accelerators, where the principles of conservation laws hold considerable importance.

In the context of collisions, specifically inelastic collisions, the momentum conservation principle holds true, ensuring that the system's total momentum remains constant. However, it is important to note that in such collisions, the conservation of kinetic energy is not fully upheld. A portion of the kinetic energy undergoes conversion into alternative forms of energy, such as thermal energy or the alteration of the colliding objects' shape.

There exist two primary categories of inelastic collisions. (a) Partially inelastic collisions involve a scenario where the colliding objects briefly adhere to each other before separating. (b) While the principle of momentum conservation holds, it is observed that the final kinetic energy is diminished in comparison to the initial kinetic energy (Wach et al., 2016).

A completely inelastic collision refers to a scenario where the objects involved adhere to one another and subsequently move as a unified entity following the collision. The objects exhibit persistent deformation or compression, resulting in minimal kinetic energy post-collision. In real-world scenarios, such as car crashes, inelastic collisions are frequently observed, wherein a portion of the energy is dissipated due to deformation and sound production.

A perfectly inelastic collision is a specific instance where the objects involved adhere to one another following the interaction and travel at a shared ultimate velocity. In the context of this collision, the entirety of the kinetic energy dissipates, resulting in the amalgamation of the colliding entities into a singular mass (Su et al., 1978).

5.5.2. Conservation of Momentum in Collisions

The derivation of the principle of conservation of momentum Equation: The principle of momentum conservation is a fundamental principle in physics, which asserts that the total momentum of a closed system remains constant in the absence of external forces. During an interaction involving a collision between two objects, the momentum of each object undergoes a change due to the forces exerted. Nevertheless, the system's overall momentum remains conserved both before and after the collision.

The mathematical expression for the conservation of momentum is as follows:

The principle of conservation of momentum states that the total initial momentum of a system is equal to the total final momentum of the system. The principle of conservation of momentum states that the total initial momentum of a system is equal to the total final momentum of the system

(Smith, 1994). In the case of a collision involving two objects, the equation is modified:

$$m_1 \cdot v_{1i} + m_2 \cdot v_{2i} = m_1 \cdot v_{1f} + m_2 \cdot v_{2f}$$

Utilizing the conservation of momentum is a highly effective method for examining collisions and anticipating the ultimate velocities of objects subsequent to the interaction. Through the conservation of momentum equation, scholars can determine velocities that are not known and gain insight into the dynamics of the collision. An illustrative scenario to examine is a basic one-dimensional elastic collision involving two objects characterized by masses m1 and m2, which are initially in motion with velocities v1i and v2i correspondingly. Following the collision, the objects exhibit final velocities denoted as v1f and v2f. The equation representing the conservation of momentum can be expressed as follows (Taguchi, 1992):

$$m_1 \cdot v_{1i} + m_2 \cdot v_{2i} = m_1 \cdot v_{1f} + m_2 \cdot v_{2f}$$

5.5.3. Coefficient of Restitution and its Significance in Collisions

The dimensionless quantity known as the coefficient of restitution (e) characterizes the elasticity of a collision. The given expression denotes the proportion between the relative velocities of the objects following the collision (v relative, after) and the relative velocities before the collision (v relative, before).

In mathematical terms, the coefficient of restitution can be expressed as:

$$e = {}_{\text{relative, after}} / {}_{\text{relative, before}}$$

The coefficient of restitution is a scalar quantity that ranges from 0 to 1, denoting various categories of collisions.

A collision with a coefficient of restitution equal to zero results in a perfectly inelastic collision where the objects involved adhere to each other and move with a shared final velocity. The resultant velocity following the collision is zero (Zhang & Vu-Quoc, 2002).

In the context of collisions, an elastic collision is characterized by the absence of any kinetic energy dissipation. The post-collision relative velocity is equivalent to the pre-collision relative velocity.

In the range of $0<e<1$, an inelastic collision occurs, wherein the objects involved separate following the collision, albeit with a loss of kinetic energy. The post-collision relative velocity is lower than the pre-collision relative velocity.

The coefficient of restitution is a significant parameter when evaluating collisions, particularly in sports and engineering. The analysis aids in comprehending the dissipation of energy and the extent of deformation that occurs during the collision (Zenit & Legendre, 2009).

5.5.4. Experimental Verification of Conservation of Momentum

The momentum conservation principle is a fundamental concept that has undergone rigorous experimental validation across various scenarios. Researchers have conducted numerous experiments to validate the conservation of momentum, focusing on collisions between objects with varying masses and velocities. In the context of a collision between two carts on a frictionless track, it is customary to measure the system's total momentum, which is the summation of the individual momenta of the two carts before the collision. Following the collision, a subsequent measurement determines the total momentum. The findings consistently indicate that the sum of momenta before the collision is equivalent to the sum after the collision, thereby confirming the momentum conservation principle. The experimental validation of the principle of conservation of momentum affirms its validity. It furnishes valuable empirical evidence for the analysis of real-world collisions and the development of safety protocols (Cao et al., 2016).

5.6. IMPULSE-MOMENTUM RELATIONSHIP AND REAL-WORLD APPLICATIONS

The impulse-momentum relationship is important in comprehending and evaluating impulsive forces within sports. Impulse is defined as the multiplication of force and time, and within the realm of sports, it frequently encompasses rapid and vigorous exchanges between athletes and their surrounding milieu. There are several notable domains within sports where the impulse-momentum relationship holds significance.

1. The Impulse-Momentum Relationship in Running and Jumping Events in Track and Field Sprinters can generate a substantial amount of impulse by exerting a significant force upon the ground during each stride, thereby propelling themselves forward. Likewise, athletes employ a substantial vertical impulse during the takeoff phase to attain increased heights or distances in the disciplines of long jump and high jump (Fang et al., 2012).

2. Ball sports, such as soccer, basketball, and volleyball, involve the application of the impulse-momentum relationship when players execute actions such as kicking, throwing, or hitting the ball. The velocity and trajectory of the ball are determined by the magnitude of the force exerted on it and the duration of the contact period. Athletes employ a strategic approach by leveraging the fundamental principle of the impulse-momentum relationship to attain forceful shots, passes, and serves.
3. Gymnastics entails utilizing the impulse-momentum relationship by gymnasts to execute a diverse range of acrobatic maneuvers. During the execution of a floor routine, gymnasts can perform flips, twists, and somersaults precisely by generating impulses through the force exerted on the ground during takeoffs.

Examining impulsive forces in sports facilitates the comprehension of movement mechanics, the enhancement of performance, and the mitigation of injury susceptibility for coaches and athletes. The optimization of applied impulse and the mitigation of harmful impulsive forces can be achieved through the implementation of appropriate training protocols and techniques in athletes (Pride et al., 1998).

5.6.1. Safety Considerations in Car Crashes and Airbag Deployment

Automobile collisions entail abrupt deceleration and substantial impulsive forces exerted upon the individuals inside the vehicle. The comprehension of impact forces encountered during collisions and the development of safety measures such as airbags and seat belts heavily rely on understanding the impulse-momentum relationship.

1. Seat belts prolong the duration of impact in the event of a collision, thereby mitigating the magnitude of force encountered by the vehicle occupants. By extending the collision duration, the force exerted on the individuals involved is distributed over a greater period, reducing the likelihood of sustaining significant injuries.
2. Airbag deployment occurs swiftly in the event of a collision, augmenting the cushioning effect. The design of airbag deployment systems is guided by the impulse-momentum relationship, which ensures rapid inflation of the airbags and optimal protection for the vehicle occupants.

3. Crumple zones are a feature found in contemporary automobiles specifically engineered to absorb the impact forces that occur during a collision effectively. Crumple zones undergo deformation upon collision, resulting in an extended duration of the impact and a consequent reduction in the impulse encountered by the individuals inside the vehicle.

A comprehensive comprehension of the impulse-momentum relationship is imperative when developing safety mechanisms that efficiently alleviate impulsive forces and safeguard individuals involved in automobile collisions (Stewart et al., 2003).

5.6.2. Impact of Force and Time on Injury Mitigation

The relationship between impulse and momentum directly influences injury mitigation in different scenarios. The magnitude of the force exerted, and the duration of the interaction are crucial factors in determining the extent of injuries. In a general context, the mitigation of injury risks can be achieved by decreasing the peak force's magnitude and prolonging the impact duration. Several examples can be provided:

1. Fall protection systems, such as safety harnesses and netting, are implemented in construction and various industries to prolong the impact duration in the event of a fall. Implementing this measure diminishes the force exerted on the worker, resulting in a decrease in the likelihood of sustaining serious injuries.
2. Sports Safety Equipment: Within contact sports such as football and hockey, athletes employ protective equipment, including helmets, pads, and mouthguards, to mitigate the risk of injury. The safety equipment in question has been specifically engineered to effectively distribute the force generated by impact and prolong the duration of interaction, thereby mitigating the impulse encountered by athletes in the event of collisions (McBurnie et al., 2022).

Did You Know?

Injury mitigation during car crashes is influenced not only by the force experienced but also by the duration of the collision. A longer collision time reduces the peak force and minimizes the risk of injuries.

5.6.3. Engineered Solutions to Reduce Impulsive Forces in Engineering

The impulse-momentum relationship holds significant importance in engineering, as it plays a pivotal role in the design of structures, machinery, and safety systems. Engineers utilize a range of methodologies to mitigate impulsive forces and augment safety measures. Several engineered solutions are available, including:

1. Shock absorbers play a crucial role in automotive and mechanical systems by mitigating vibrations and minimizing the impact force exerted on the system. These devices operate by converting the energy associated with motion into thermal energy, prolonging the impact duration and diminishing the magnitude of force transferred to the structure or the individuals present (Knapik et al., 2014).
2. Energy-absorbing Inimpact-resistant structures, engineers employ energy-absorbing materials such as foam or crumple zones to mitigate the impulse encountered during collisions. Upon impact, these materials undergo deformation, resulting in an extended collision duration and a gradual dissipation of energy.
3. Vibration isolation systems are commonly utilized in machinery and industrial equipment to mitigate the propagation of impulsive forces to the surrounding environment. These systems serve the purpose of safeguarding fragile instruments and structures against potential harm resulting from impulsive forces (Eager et al., 2016).

5.7. ANGULAR MOMENTUM AND ROTATIONAL MOTION

Angular momentum is a fundamental principle in physics that characterizes an object's rotational dynamics with respect to a stationary axis. The concept is comparable to linear momentum, although it pertains to rotational quantities rather than linear ones. The concept of angular momentum (L) can be mathematically expressed as the multiplication of the moment of inertia (I) and the angular velocity (ω) of the object:

$$L=I\cdot\omega$$

The symbol "L" represents angular momentum, which is typically measured in units of kilogram meter squared per second (kg m²/s) or joule seconds (J s). The symbol "I" represents the object's moment of inertia, measured in kilograms multiplied by meters squared (kg m²). The symbol ω represents the angular velocity of the object, measured in radians per second (rad/s). The unit of angular momentum is commonly expressed as kilogram-meter squared per second (kg m²/s) or joule-second (J s), which is considered equivalent to the unit of linear momentum. Angular momentum is a vector quantity that possesses both magnitude and direction. Determining the direction of angular momentum follows the right-hand rule, wherein the thumb denotes the axis of rotation's direction, while the curled fingers represent the direction of angular velocity (Dixon, 1970).

5.7.1. Conservation of Angular Momentum

The principle of conservation of angular momentum is a fundamental concept in physics, which asserts that the overall angular momentum of a closed system remains constant in the absence of any external torques acting upon the said system. The principle above resembles conserving linear momentum when no external forces act upon the system.

Mathematically, the principle of conservation of angular momentum can be formally expressed in the following manner:

Total initial angular momentum=Total final angular momentum

Or,

$$I_1 \cdot \omega_1 + I_2 \cdot \omega_2 + \ldots + I_n \cdot \omega_n = I_1 \cdot \omega_1' + I_2 \cdot \omega_2' + \ldots + I_n \cdot \omega_n'$$

The principle of angular momentum conservation is a fundamental concept widely utilized in diverse contexts about rotational motion, encompassing celestial mechanics, spinning objects, and figure skating (Bliokh & Bliokh, 2006).

5.7.2. Torque and its Relationship with Angular Momentum

Torque can be understood as the rotational counterpart to force in linear motion. The concept referred to is a force's rotational tendency to induce angular motion in an object about a specific axis.

The interdependence among torque (τ), angular momentum (L), and angular acceleration (α) is elucidated by the subsequent equation (Wang et al., 2006):

$$\tau = I \cdot \alpha$$

where: τ = Torque (N m or J); I = Moment of inertia of the object (kg m²); α = Angular acceleration (rad/s²).

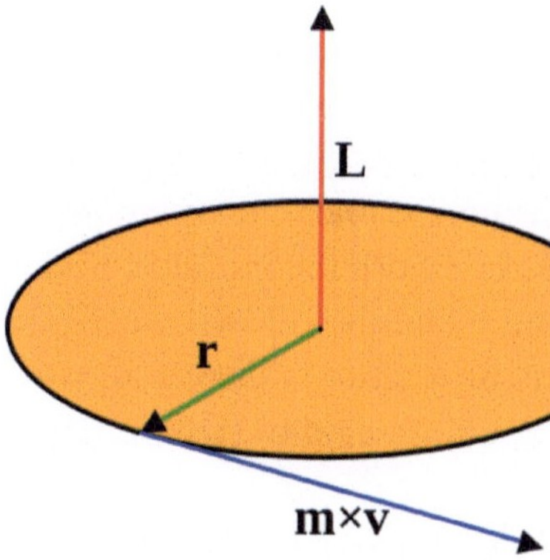

Figure 5.4. Relation between torque and angular momentum (Source: Toppr, Creative Commons License)

The torque in this equation equals the multiplication of the moment of inertia and the angular acceleration (Figure 5.4). The moment of inertia pertains to an object's inherent capacity to resist alterations in its rotational motion. This is analogous to how mass characterizes an object's resistance to modifications in its linear motion. Angular acceleration is the derivative of angular velocity concerning time and is induced by torque exertion.

The elucidation of the connection between torque and angular momentum can be enhanced by examining the temporal derivative of the equation governing angular momentum:

$dt/dL=d(I\cdot\omega)/dt=I\cdot d\omega/dt$

Using the definition of angular acceleration ($\alpha=d\omega/dt$), we find:

$dL/dtm=I\cdot\alpha=\tau$

The equation above demonstrates that the angular momentum alteration rate is equivalent to the exerted torque. Therefore, applying torque to an object results in a modification of its angular momentum (Kajita et al., 2001).

5.7.3. Application of Angular Momentum in Engineering and Physics

The concept of angular momentum finds extensive utility in engineering and physics, playing a crucial role in elucidating the principles governing rotational motion and facilitating the development of diverse mechanical systems. Several practical applications can be attributed to it.

1. Gyroscopes are mechanical devices that leverage the principles of angular momentum to uphold stability and orientation. They are extensively employed in various applications, such as navigation systems, spacecraft, and stabilization mechanisms in vehicles and ships. Gyroscopes utilize the principle of angular momentum conservation to resist alterations in orientation and uphold a constant axis of rotation (Kajita et al., 2001).
2. The role of angular momentum is of utmost importance in sports activities that entail rotational or spinning motions. Figure skaters, divers, and gymnasts employ the fundamental concept of angular momentum to perform rotations, aerial maneuvers, and contortions. Athletes can manipulate their moments of inertia and angular velocities through deliberate control of their body positions. This enables them to execute a diverse range of acrobatic movements.
3. Celestial mechanics relies heavily on angular momentum conservation to comprehend the dynamics of celestial bodies such as planets, moons, and other astronomical entities. This concept can be illustrated in Kepler's laws of planetary motion, which elucidate the trajectories of planets revolving around the sun, and are contingent upon the principle of angular momentum conservation.
4. Flywheels are devices used for energy storage that operate based on the principles of angular momentum. Rotational energy is stored within these mechanisms upon torque application and discharged as required. Flywheels find utility in various domains, including energy storage systems, mechanical engineering, and automotive technology.
5. The study of rotational dynamics in machines is a topic of interest in engineering. Comprehending angular momentum is important in analyzing motion in machines and mechanical systems, particularly about rotating components such as gears, wheels, and turbines. Engineers apply principles related to angular momentum to enhance said machines' design and operational effectiveness (Catelan & Theuns, 1996).

5.8. EXPERIMENTAL TECHNIQUES AND MEASUREMENTS

5.8.1. Methods for Measuring Momentum and Impulse

The measurement of momentum and impulse holds significant importance in experimental physics as it enables a comprehensive understanding of object motion, facilitates the analysis of collisions, and provides validation for theoretical models. Various methodologies and instruments are employed to quantify these variables precisely. Several commonly used techniques for quantifying momentum and impulse are as follows:

1. Motion tracking systems employ cameras and markers to monitor objects' spatial position and movement in a three-dimensional environment. By capturing the path followed by an object both prior to and subsequent to a collision, researchers can ascertain the alteration in momentum and subsequently compute the impulse encountered during the interaction (Smith et al., 2008).
2. Force sensors, such as strain gauges or load cells, quantify the magnitudes of forces exerted in collisions or interactions. The impulse experienced by an object can be determined by integrating the measured force over a given period.
3. Pendulum experiments are frequently employed in the measurement of momentum and impulse. Researchers can analyze the impulse by observing the motion of a pendulum with a known mass and initial velocity after it collides with another object, thereby determining the change in momentum.
4. The ballistic pendulum is a scientific apparatus employed to determine the velocity of a projectile. The apparatus comprises a pendulum apparatus, wherein a target is suspended at the lower end. The conservation of momentum is observed when the projectile makes contact with the target and becomes embedded within it. Through the utilization of height measurements attained by the pendulum subsequent to the collision event, researchers can ascertain the projectile's initial velocity and subsequently compute its momentum and impulse.
5. The experiments on linear and angular momentum conservation entail systematically observing object motion within controlled environments, such as an air table or rotational apparatus.

Researchers can accurately ascertain momentum and impulse in different scenarios by employing momentum conservation principles and measuring pertinent parameters (Rigby et al., 2019).

5.8.2. Collisions in the Laboratory Setting

Laboratory collisions are paramount in experimental physics because they facilitate the examination of momentum and energy conservation, verify theoretical frameworks, and investigate the dynamics of diverse materials and objects during interactions. Various collision types are examined within laboratory settings, encompassing elastic, inelastic, and partially inelastic collisions. Various methodologies and configurations are utilized to carry out these experiments (Nicholls et al., 2007).

1. Projectile collisions refer to the experimental process of propelling objects, known as projectiles, towards a designated target and subsequently analyzing the ensuing interactions. In the context of a ballistics experiment, researchers can investigate the trajectory of bullets or pellets propelled at different velocities and angles, ultimately impacting a target composed of a specific material. By examining the ultimate velocities and deformations exhibited by the projectiles and targets, researchers can acquire valuable knowledge pertaining to the transfer of momentum and the dissipation of energy.
2. Rotational collisions involve the interaction between objects possessing angular momentum. As an illustration, scholars may investigate spinning tops or rotating disks colliding, primarily emphasizing preserving angular momentum. These experiments facilitate the comprehension of the dynamics exhibited by rotating objects and serve to validate the fundamental principles governing the conservation of angular momentum (Velarde et al., 2012).
3. Particle colliders are sophisticated experimental configurations utilized in high-energy physics to investigate subatomic particles. Within these colliders, particles are subjected to acceleration, resulting in their attainment of high velocities. Subsequently, these particles are brought into collision, achieving velocities that approach the speed of light. The resultant collisions yield

significant data regarding the interactions between particles and the fundamental forces that govern the universe.

4. The examination of drop tests and crash tests. In materials testing and engineering, researchers and professionals conduct drop and crash tests to evaluate the impact resistance and safety characteristics of various materials and structures. During crash tests, automobiles or their components are exposed to controlled collisions to evaluate their performance in realistic situations. These experiments yield significant data that can be utilized in developing safer vehicles and evaluating the impulse encountered during collisions (Porter et al., 2010).

5.8.3. High-Energy Physics Experiments

Large-scale particle accelerators and colliders, such as the renowned Large Hadron Collider (LHC) at the European Organization for Nuclear Research (CERN), serve as the primary venues for the execution of high-energy physics experiments. The conducted experiments entail the acceleration of subatomic particles, specifically protons or electrons, to velocities of significant magnitude, followed by their collision at energies that closely approach the speed of light. The collisions give rise to events of high energy, which are examined through the utilization of advanced detectors and techniques for data analysis. The objectives of experiments in the field of high-energy physics encompass (Lecoq et al., 1992):

1. The process of uncovering novel particles involves the occurrence of high-energy collisions, which have the potential to generate particles that are not commonly observed in ordinary interactions. By examining these collisions and identifying novel particles, researchers can acquire valuable knowledge regarding the fundamental building blocks of matter and the underlying forces that dictate their interactions.
2. Investigating the Essence of Forces: Using high-energy collisions offers an avenue for delving into the fundamental forces that govern the natural world, including the strong nuclear force, weak nuclear force, and electromagnetism. By observing the interactions between particles during collisions, researchers can enhance their comprehension of the fundamental characteristics and properties of these forces (Zhou et al., 2021).
3. High-energy physics experiments provide a means for researchers

to empirically evaluate predictions using fundamental theories, such as the Standard Model of particle physics. Physicists can assess the validity of theoretical models and explore novel physics beyond the existing comprehension by juxtaposing empirical findings with theoretical projections (Chekanov, 2015).

4. Investigating the Early Universe: High-energy collisions replicate the conditions that closely resemble those observed in the nascent universe, occurring shortly after the Big Bang. By examining these collisions, researchers can acquire valuable knowledge about the initial phases of the universe and its subsequent development over time (Romaniuk & Pozniak, 2007).

The utilization of experimental methodologies to quantify momentum and impulse plays a pivotal role in validating theoretical frameworks and comprehending object motion. Various techniques, including motion tracking systems, force sensors, and pendulum experiments, are utilized to quantify these variables precisely. The laboratory environment is utilized to investigate collisions to analyze the transfer of momentum, dissipation of energy, and the behavior exhibited by various materials and objects during their interactions (Wang et al., 2016). Collisions encompass a spectrum of phenomena, spanning from elementary projectile and rotational collisions to intricate particle colliders employed in experiments in high-energy physics. In the field of high-energy physics, scientists engage in the process of accelerating subatomic particles to velocities that approach the speed of light. By analyzing the resulting collisions, these researchers aim to understand better fundamental particles, forces, and the universe's origins (Lindström et al., 1999). Using experimental methodologies and sophisticated instrumentation enables researchers to investigate the complexities associated with momentum, impulse, and the dynamics of matter and energy across different magnitudes, thereby facilitating a more profound comprehension of natural phenomena (Chen et al., 2004).

Activity 1.1

Imagine two cars of different masses are moving towards each other. Car A has a mass of 1000 kg and is moving at 20 m/s, while Car B has a mass of 1500 kg and is moving at 15 m/s in the opposite direction. What is the total momentum of the system before and after the collision, assuming it is perfectly inelastic? Calculate the change in momentum and the impulse experienced during the collision.

SUMMARY

This chapter introduces the concepts of momentum and impulse in the context of collisions. It explains how momentum and impulse are related and explores the conservation of linear momentum in elastic and inelastic collisions. The chapter also delves into the significance of impulse-momentum relationships in real-world applications, such as car crashes and engineering design. Additionally, it covers angular momentum and its conservation in rotational motion, along with experimental techniques used to measure momentum and impulse in various scenarios.

REVIEW QUESTIONS

1. What is the relationship between impulse and momentum?
2. Define linear momentum and explain its conservation in a closed system.
3. Differentiate between elastic and inelastic collisions, focusing on kinetic energy conservation.
4. How do engineers utilize impulse-momentum relationships to design safety features like airbags?
5. Explain the coefficient of restitution and its role in characterizing collisions.
6. Discuss the importance of the duration of a collision in mitigating injury forces during car crashes.
7. What is angular momentum, and how is it conserved in rotational motion?
8. Describe some experimental techniques used to measure momentum and impulse in laboratory settings.

REFERENCES

1. Anders, A. (2011). Discharge physics of high power impulse magnetron sputtering. *Surface and Coatings Technology, 205*, S1–S9.
2. Belinfante, F. J. (1940). On the current and the density of the electric charge, the energy, the linear momentum and the angular momentum of arbitrary fields. *Physica, 7*(5), 449–474.
3. Bliokh, K. Y., & Bliokh, Y. P. (2006). Conservation of angular momentum, transverse shift, and spin Hall effect in reflection and refraction of an electromagnetic wave packet. *Physical Review Letters, 96*(7), 073903.
4. Cao, D., Yang, Y., Chen, H., Wang, D., Jiang, G., Li, C., ... & Zhao, K. (2016). A novel contact force model for the impact analysis of structures with coating and its experimental verification. *Mechanical Systems and Signal Processing, 70*, 1056–1072.
5. Catelan, P., & Theuns, T. (1996). Evolution of the angular momentum of protogalaxies from tidal torques: Zel'dovich approximation. *Monthly Notices of the Royal Astronomical Society, 282*(2), 436–454.
6. Chekanov, S. V. (2015). HepSim: a repository with predictions for high-energy physics experiments. *Advances in High Energy Physics, 2015*.
7. Chen, J., Zhang, L., & Zhu, R. Y. (2004). Large size LYSO crystals for future high energy physics experiments. In *IEEE Symposium Conference Record Nuclear Science 2004*. (Vol. 1, pp. 117–125). IEEE.
8. Dehong, Y., Guangxi, D., Qingzheng, L., Zhiyong, H., & Yujin, Q. (1994). Application of the Modified BUU Model to Reaction Induced by 16 O on 197 An at 20MeV≤ E/u≤ 180 MeV: Dependence of the Angular and Linear Momentum Transfer on Impact Parameter and Energy. *Chinese Physics C, 18*(11), 1035–1042.
9. Dixon, W. G. (1970). Dynamics of extended bodies in general relativity. I. Momentum and angular momentum. *Proceedings of the Royal Society of London. A. Mathematical and Physical Sciences, 314*(1519), 499–527.
10. Duthinh, D., & Simiu, E. (2010). Safety of structures in strong winds and earthquakes: multihazard considerations. *Journal of Structural Engineering, 136*(3), 330–333.
11. Eager, D., Hayati, H., & Chapman, C. (2016). Impulse force as an additional safety criterion for improving the injury prevention

performance of impact attenuation surfaces in children's playgrounds. In *ASME International Mechanical Engineering Congress and Exposition* (Vol. 50688, p. V014T14A030). American Society of Mechanical Engineers.

12. Engelhardt, A. G., & Phelps, A. V. (1963). Elastic and inelastic collision cross sections in hydrogen and deuterium from transport coefficients. *Physical Review, 131*(5), 2115.
13. Fang, N., Nielson, K., & Kawamura, S. M. (2012). Integrating Real World Engineering Examples and Mathematical Calculations Into Computer Simulations to Improve Students' Understanding of Concept Pairs. In *2012 ASEE Annual Conference & Exposition* (pp. 25–805).
14. Fukui, H., Fujitani, H., Mukai, Y., Ito, M., & Mosqueda, G. (2021). Response evaluation and analysis using impulse of base-isolated buildings during a collision with retaining wall. *Japan Architectural Review, 4*(1), 88–104.
15. Han, I. (2015). Impulse-momentum based analysis of vehicle collision accidents using Monte Carlo simulation methods. *International Journal of Automotive Technology, 16*, 253–270.
16. Hnizdo, V. (1992). Conservation of linear and angular momentum and the interaction of a moving charge with a magnetic dipole. *American Journal of Physics, 60*(3), 242–246.
17. Kajita, S., Yokoi, K., Saigo, M., & Tanie, K. (2001). Balancing a humanoid robot using backdrive concerned torque control and direct angular momentum feedback. In *Proceedings 2001 ICRA. IEEE International Conference on Robotics and Automation (Cat. No. 01CH37164)* (Vol. 4, pp. 3376–3382). IEEE.
18. Kim, B. K., Cho, J. U., Jeong, K. Y., Chun, Y., Carman, G. P., & Cheon, S. S. (2019). A coupled constitutive relation with impulse-momentum for compressive impact behavior of the expanded polypropylene foam. *Polymer Engineering & Science, 59*(1), 49–57.
19. Knapik, J. J., Trone, D. W., Tchandja, J., & Jones, B. H. (2014). Injury-reduction effectiveness of prescribing running shoes on the basis of foot arch height: summary of military investigations. *journal of Orthopaedic & Sports Physical Therapy, 44*(10), 805–812.
20. Kuhl, D., & Ramm, E. (1996). Constraint energy momentum algorithm and its application to non-linear dynamics of shells. *Computer Methods in Applied Mechanics and Engineering, 136*(3–4), 293–315.

21. Lecoq, P., Schussler, M., & Schneegans, M. (1992). Progress and prospects in the development of new scintillators for future high energy physics experiments. *Nuclear Instruments and Methods in Physics Research Section A: Accelerators, Spectrometers, Detectors and Associated Equipment*, *315*(1–3), 337–343.
22. Lindström, G., Moll, M., & Fretwurst, E. (1999). Radiation hardness of silicon detectors–a challenge from high-energy physics. *Nuclear Instruments and Methods in Physics Research Section A: Accelerators, Spectrometers, Detectors and Associated Equipment*, *426*(1), 1–15.
23. McBurnie, A. J., Harper, D. J., Jones, P. A., & Dos' Santos, T. (2022). Deceleration training in team sports: Another potential 'vaccine'for sports-related injury?. *Sports Medicine*, 1–12.
24. McGuire, P., & Kouri, D. J. (1974). Quantum mechanical close coupling approach to molecular collisions. jz-conserving coupled states approximation. *The Journal of Chemical Physics*, *60*(6), 2488–2499.
25. Meyers, J., & Meneveau, C. (2013). Flow visualization using momentum and energy transport tubes and applications to turbulent flow in wind farms. *Journal of Fluid Mechanics*, *715*, 335–358.
26. Nicholls, M. E., Loftus, A., Mayer, K., & Mattingley, J. B. (2007). Things that go bump in the right: The effect of unimanual activity on rightward collisions. *Neuropsychologia*, *45*(5), 1122–1126.
27. Porter, C. O., Webb, J. W., & Gogus, C. I. (2010). When goal orientations collide: Effects of learning and performance orientation on team adaptability in response to workload imbalance. *Journal of Applied Psychology*, *95*(5), 935.
28. Pride, T. O. B., Vokos, S., & McDermott, L. C. (1998). The challenge of matching learning assessments to teaching goals: An example from the work-energy and impulse-momentum theorems. *American Journal of Physics*, *66*(2), 147–157.
29. Rigby, S. E., Tyas, A., Curry, R. J., & Langdon, G. S. (2019). Experimental measurement of specific impulse distribution and transient deformation of plates subjected to near-field explosive blasts. *Experimental Mechanics*, *59*, 163–178.
30. Romaniuk, R. S., & Pozniak, K. T. (2007). Metrological aspects of accelerator technology and high energy physics experiments. *Measurement Science and Technology*, *18*(8), E01.

31. Scalora, M., D'Aguanno, G., Mattiucci, N., Bloemer, M. J., Centini, M., Sibilia, C., & Haus, J. W. (2006). Radiation pressure of light pulses and conservation of linear momentum in dispersive media. *Physical Review E, 73*(5), 056604.
32. Schilling, B. K., Falvo, M. J., & Chiu, L. Z. (2008). Force-velocity, impulse-momentum relationships: Implications for efficacy of purposefully slow resistance training. *Journal of Sports Science & Medicine*, *7*(2), 299.
33. Schilling, B. K., Falvo, M. J., & Chiu, L. Z. (2008). Force-velocity, impulse-momentum relationships: Implications for efficacy of purposefully slow resistance training. *Journal of Sports Science & Medicine*, *7*(2), 299.
34. Smith, G. C. (1994). *Conservation of Momentum Analysis of Two-Dimensional Colliding Bodies, with or Without Trailers* (No. 940566). SAE Technical Paper.
35. Smith, P., Mostert, F., & Snyman, I. (2008). Comparison of methods to measure the blast impulse loading of an explosive charge. In *24th International Symposium on Ballistics, New Orleans* (pp. 1226–1234).
36. Stewart, T. C., Girotti, M. J., Nikore, V., & Williamson, J. (2003). Effect of airbag deployment on head injuries in severe passenger motor vehicle crashes in Ontario, Canada. *Journal of Trauma and Acute Care Surgery*, *54*(2), 266–272.
37. Su, T., Su, E. C., & Bowers, M. T. (1978). Ion–polar molecule collisions. Conservation of angular momentum in the average dipole orientation theory. The AADO theory. *The Journal of Chemical Physics*, *69*(5), 2243–2250.
38. Taguchi, M. (1992). A method for calculating neoclassical transport coefficients with momentum conserving collision operator. *Physics of Fluids B: Plasma Physics*, *4*(11), 3638–3643.
39. Triyani, G., Danawan, A., Suyana, I., & Kaniawati, I. (2019). An investigation of students' misconceptions about momentum and impulse through interactive conceptual Instruction (ICI) with computer simulation. In *Journal of Physics: Conference Series* (Vol. 1280, No. 5, p. 052008). IOP Publishing.
40. Velarde, J. J., Alvarado, C., & Mijangos, A. (2012). Evaluating an interactive lecture demonstration implementation in a lab setting: An example from a collisions and momentum learning activity. *Lat. Am. J. Phys. Educ. Vol*, *1*(192), 6.

41. Wach, W., Gidlewski, M., & Prochowski, L. (2016). Modelling reliability of vehicle collision reconstruction based on the law of conservation of momentum and Burg equations. In *20th International Scientific Conference Transport Means* (pp. 693–698).
42. Wang, C. J., Liu, Z. A., Zhao, J. Z., & Liu, Z. (2016). Design of a high throughput electronics module for high energy physics experiments. *Chinese Physics C*, *40*(6), 066102.
43. Wang, Y., Xia, K., Su, Z. B., & Ma, Z. (2006). Consistency in formulation of spin current and torque associated with a variance of angular momentum. *Physical Review Letters*, *96*(6), 066601.
44. Xie, X. Y., & Liu, X. B. (2020). Elastic and inelastic collisions of the semirational solutions for the coupled Hirota equations in a birefringent fiber. *Applied Mathematics Letters*, *105*, 106291.
45. Zenit, R., & Legendre, D. (2009). The coefficient of restitution for air bubbles colliding against solid walls in viscous liquids. *Physics of Fluids*, *21*(8).
46. Zhang, X., & Vu-Quoc, L. (2002). Modeling the dependence of the coefficient of restitution on the impact velocity in elasto-plastic collisions. *International Journal of Impact Engineering*, *27*(3), 317–341.
47. Zhou, X., Du, S., Li, G., & Shen, C. (2021). TopoAna: A generic tool for the event type analysis of inclusive Monte-Carlo samples in high energy physics experiments. *Computer Physics Communications*, *258*, 107540.
48. Zohar, D. (1980). Safety climate in industrial organizations: theoretical and applied implications. *Journal of Applied Psychology*, *65*(1), 96.

CHAPTER 6

Rotational Motion

Contents

6.1. INTRODUCTION

Rotational motion is a basic principle in physics that pertains to the motion exhibited by an object as it rotates around a specific axis. In contrast to translational motion, which involves the linear movement of an object along a straight path, rotational motion refers to the circular or spinning motion of an object around a stationary axis. This particular form of motion is widely observed in diverse natural phenomena and human-engineered systems, and its comprehension is essential for elucidating the dynamics of rotating objects, celestial entities, and mechanical apparatus (Bishop, 1990).

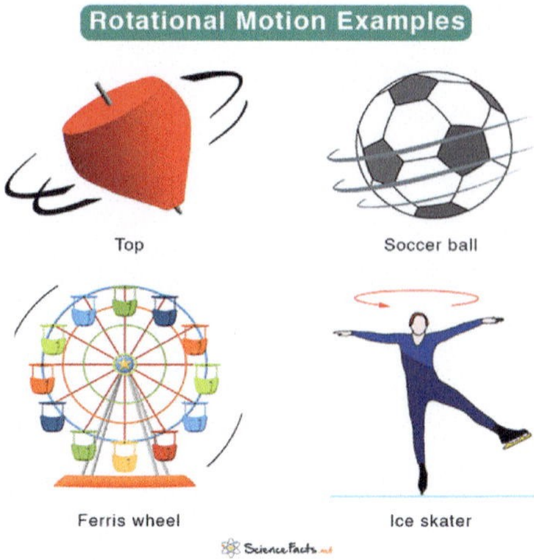

Figure 6.1. Examples of rotational motion (Source: Science Facts, Creative Commons License)

Rotational motion can be characterized as the translational movement of an entity along a circular trajectory revolving around a central axis. The axis may be situated either internally or externally about the object. When an object undergoes rotational motion, various points on the object traverse circular trajectories with distinct radii. Nevertheless, it is worth noting that all of these points exhibit a shared angular velocity, which measures the object's rotational speed around the axis (Figure 6.1).

Rotational motion encompasses several key attributes: angular velocity, angular acceleration, and angular displacement. Angular displacement, denoted as θ, refers to the magnitude of the angle by which an object rotates. This measurement can be expressed in either radians or degrees. Angular velocity (ω) refers to the temporal variation of angular displacement and is quantified in radians per second (rad/s) units. Angular acceleration (α) is the temporal variation of angular velocity and is quantified in units of radians per second squared (Cochard et al., 2006).

6.2. COMPARISON WITH TRANSLATIONAL MOTION

In order to gain a comprehensive understanding of rotational motion, it is imperative to undertake a comparative analysis with translational motion, which pertains to the linear displacement of an object between two distinct positions. In translational motion, an entity undergoes movement along a linear path, and its location is characterized using linear displacement, velocity, and acceleration.

Although translational and rotational motions both entail alterations in position over time, they exhibit distinct fundamental quantities and characteristics. In the context of translational motion, the quantities of displacement, velocity, and acceleration are vector quantities that characterize the alterations in position, speed, and direction of an object as it moves along a linear trajectory (Figure 6.2) (Hadziioannou et al., 2012).

One notable distinction between the two forms of motion is the utilization of Newton's laws. The relationship between the net force acting on an object and its linear acceleration in translational motion is described by Newton's second law (F=ma). In the context of rotational motion, there is a corresponding relationship between the net torque (τ) and the angular acceleration (α) as described by the rotational equivalent of Newton's second law ($\tau = I \cdot \alpha$), where I represent the moment of inertia of the object (Frobin et al., 2002).

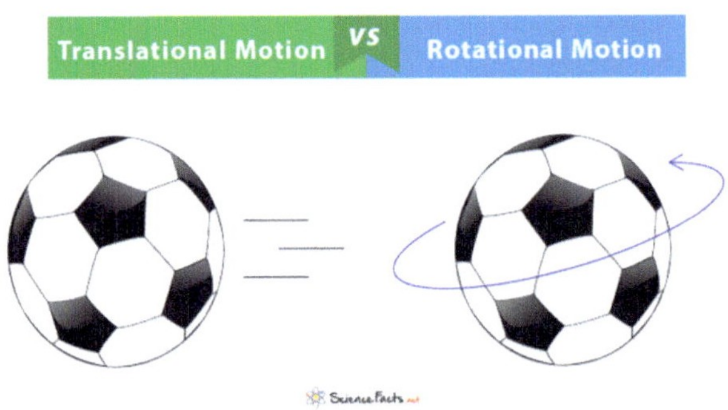

Figure 6.2. Comparison of rotational vs. translational motion (Source: Science Facts, Creative Commons License)

6.3. ROTATIONAL KINEMATICS

The field of rotational kinematics pertains to motion characterization by utilizing angles and angular quantities. In the context of rotational motion, an object undergoes a trajectory that follows a circular path centered around an axis of rotation. Angular displacement, denoted as θ, is a quantitative representation of the angle at which an object rotates around a specific axis. The measurement of an object's orientation, expressed in either radians or degrees, is defined as the change in its orientation.

The following formula can express the relationship between angular displacement and linear displacement (s):

$$s = r \cdot \theta$$

The variable "r" represents the radial distance, which is the distance from the axis of rotation to a specific point on the object. In the case of angular displacements that are relatively small, it is possible to make an approximation of the arc length, denoted as s, using the equation $s \approx r \cdot \theta$. This approximation is based on the observation that the arc length and the radius of the circle can be considered as the two sides of a small triangle (Mashood & Singh, 2012).

Angular velocity (ω) refers to the rate at which angular displacement changes over time. The angular velocity measures the rate at which an object rotates about an axis, expressed in radians per second (rad/s). In mathematical terms, the expression for angular velocity is defined as:

$$\omega = d\theta/dt$$

The symbol $d\theta/dt$ denotes the temporal rate of change of angular displacement, denoted by θ. Similarly, angular acceleration (α) can be defined as the temporal derivative of angular velocity. The measured quantity is the rate at which the object's angular velocity changes, denoted in units of radians per second squared (rad/s^2). In mathematical terms, the expression for angular acceleration is (Hajiaghamemar et al., 2020):

$$\alpha = d\omega/dt = d^2\theta/d^2t$$

6.3.1. Angular vs. Linear Quantities

The interrelationship between angular and linear quantities in rotational motion facilitates the conversion of measurements between these two systems. Let us contemplate a specific location on an object undergoing rotational motion, which is positioned at a radial distance denoted as r from the axis of rotation. If a point moves with a linear velocity v along the circumference of a circle with radius r, the corresponding angular velocity ω can be determined using the following equation:

$\omega = rv$

This relationship demonstrates that the linear velocity of a point located on a rotating object is directly proportional to the radial distance between the point and the axis of rotation. In a similar vein, the relationship between angular acceleration (α) and linear acceleration (a) can be expressed as follows:

$a = r \cdot \alpha$

Again, this equation indicates that linear acceleration is directly proportional to the radial distance from the axis of rotation (Kolyada et al., 2010).

6.3.2. Rotational Kinematic Equations and Applications

The rotational kinematic equations resemble the linear ones and offer significant insights into the dynamics of rotating entities. The rotational kinematic equations presented below can be derived for a system with constant angular acceleration.

Angular Displacement Equation: $\theta = \theta_0 + \omega_0 \cdot t + \cdot \alpha \cdot t^2$

where θ_0 is the initial angular displacement, ω_0 is the initial angular velocity, α is the angular acceleration, and *t* is the time elapsed.

Final Angular Velocity Equation: $\omega = \omega_0 + \alpha \cdot t$

where ω is the final angular velocity.

Angular Displacement and Final Angular Velocity Equation: $\omega^2 = \omega_0^2 + 2 \cdot \alpha \cdot \theta$

where θ is the angular displacement.

The equations above possess significant utility in examining the rotational motion of objects subject to a consistent angular acceleration. Angular velocity, angular displacement, and the time needed for an object to achieve a specific angular velocity or angular displacement can be determined using these equations (Zhang et al., 2018).

6.3.3. Angular and Tangential Acceleration in Circular Motion

In the context of circular motion, an object traverses a circular path revolving around a central axis of rotation. This phenomenon gives rise to two distinct forms of acceleration: angular acceleration (α) and tangential acceleration (at). Angular acceleration (α) is the temporal variation of angular velocity (ω). The entity in question is accountable for altering the object's rotational velocity and orientation. When an object undergoes circular motion at a consistent velocity, it encounters centripetal acceleration (ac), oriented toward the circle's center. The relationship between centripetal acceleration and angular velocity (ω), and linear velocity (v) can be expressed by the following equation:

$$a_c = r \cdot \omega^2 = r \cdot v^2/r^2 = v^2/r$$

The tangential acceleration (at) refers to the acceleration component aligned with the tangent of the circular path. It is responsible for alterations in the linear velocity. The equation for the tangential acceleration (at) of an object undergoing constant angular acceleration (α) while moving in a circular path with a constant radius (r) is given by the formula (Leclercq et al., 2013):

$$at = r \cdot \alpha.$$

In the context of uniform circular motion, characterized by a constant speed but changing direction, it is observed that the tangential acceleration is negligible. In contrast, the centripetal acceleration remains the sole contributing factor. Comprehending circular motion necessitates a comprehensive understanding of both angular and tangential acceleration, which serve as fundamental concepts in numerous practical contexts. These contexts include planets' orbital motion, rotating machinery's operation, and amusement park rides' dynamics (Nutsathaporn et al., 2023).

6.4. MOMENT OF INERTIA AND ROTATIONAL DYNAMICS

The concept of moment of inertia (Figure 6.3), represented by the symbol I, is a fundamental characteristic within the field of rotational dynamics. It measures an object's inherent opposition to alterations in its rotational motion. In the context of translational motion, it can be observed that it fulfills a comparable function to mass. In this regard, mass serves as a metric for quantifying an object's opposition to alterations in linear motion. Within the realm of rotational dynamics, the moment of inertia serves as a measure of an object's rotational mass, elucidating how its mass is distributed about the axis of rotation (Mulhayatiah et al., 2018).

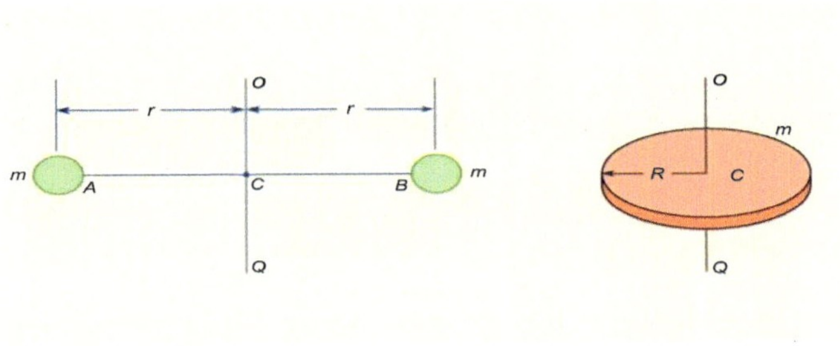

Figure 6.3. Moment of Inertia (Source: Britannica, Creative Commons License)

The moment of inertia for a point mass can be determined by the formula $I = m \cdot r^2$, where m represents the mass of the point mass, and r denotes the perpendicular distance from the axis of rotation to the point mass. Nevertheless, when dealing with objects of larger size or systems composed of multiple particles, the computation of the moment of inertia becomes considerably more intricate (Bills & James, 1999).

6.4.1. Calculation of Moment of Inertia for Different Shapes

The moment of inertia depends on how heavy an item is and how its mass is spread around its axis of rotation. Different methods can be used to determine the moment of inertia for different geometric shapes.

1. In the case of a rod rotating about its center: Let us contemplate a slender rod characterized by a length L and a mass m, engaged in rotational motion around an axis perpendicular to the rod and passes through its center. The equation representing the moment of inertia for a rod of this nature is expressed as $I = 1/2mL^2$.
2. In the case of a rod rotating about a fixed axis perpendicular to the rod through one of its ends, there is a variation in the moment of inertia. The expression for the moment of inertia of a rod rotating about one end is given by $I = (1/3)mL^2$.
3. In the case of a disc rotating about its center, the moment of inertia (I) can be determined for a disc, also known as a circular plate, with a radius (R) and mass (m). The disc rotates about an axis perpendicular to the disc and passes through its center. The moment of inertia is calculated as $I = 1/2mR^2$ (Rani, 2021).
4. In the case of a hoop, which refers to a thin circular ring with a radius denoted as R and a mass represented by m, undergoing rotation around an axis that is perpendicular to the hoop and

passes through its center, the moment of inertia can be determined as $I = mR^2$.

5. In the case of a sphere rotating about its center: The moment of inertia, denoted as I, for a solid sphere with a radius R and mass m, rotating about an axis passing through its center, is given by the equation $I = (2/5)\, mR^2$.
6. In the case of a cylinder rotating about its center: The moment of inertia (I) for a solid cylinder with a radius (R) and mass (m) rotating about an axis perpendicular to the cylinder through its center can be calculated using the formula $I = 1/2mR^2$.
7. In a solid sphere undergoing rolling motion without slipping, the moment of inertia encompasses the combined effects of both rotational and translational motion. The moment of inertia, denoted as I, for a solid sphere with a radius R and mass m, which rolls without slipping on a flat surface, is given by the equation $I = (2/5)mR^2$.

It is imperative to acknowledge that the moment of inertia is consistently a positive scalar quantity contingent upon the axis of rotation (Damavandi et al., 2009).

6.4.2. Torque and Angular Acceleration

In the context of translational motion, Newton's second law postulates that the total force exerted on an object is directly proportional to its linear acceleration. The equation F = m·a represents the relationship between the net force (F), the mass of an object (m), and its linear acceleration (a). In the field of rotational dynamics, the application of Newton's second law is extended to incorporate torque (τ) and angular acceleration (α), which serve as the rotational counterparts to force and linear acceleration, respectively.

$$\tau = I \cdot \alpha$$

Torque can be defined as a quantitative measure of the efficacy of a force in generating rotational motion. The result depends not only on the amount of force applied but also on the length of the lever arm, which is the distance between the axis of rotation and the force's line of action. This expression is used to figure out torque:

$$\tau = r \cdot F \cdot \sin(\theta),$$

where r = length of the lever arm, and F = force being applied.

Angular acceleration measures the rate at which the angular velocity of an object varies over time (Hasan & Hamarash, 2017).

Tip

When calculating angular acceleration using the torque equation, ensure that the moment of inertia is appropriately considered for the specific shape of the rotating object.

6.4.3. Rotational Equilibrium and Center of Mass

Rotational equilibrium is achieved when an object remains stationary or rotates at a consistent angular velocity. In order for an object to achieve rotational equilibrium, the sum of all torques acting upon it must be equal to zero. Mathematically, the condition above can be formally represented as the summation of torques ($\sum\tau$) equal to zero. Here, $\sum\tau$ denotes the collective sum of all torques exerted on the object with respect to a particular axis of rotation. In the state of rotational equilibrium, the object undergoes no angular acceleration, resulting in a constant angular velocity.

The centroid of an object is a singular point that symbolizes the calculated average position of all constituent mass elements. In the context of a rigid body, it is observed that the center of mass remains stationary throughout the rotational motion. The point referred to is the axis around which the object experiences exclusively rotational motion, resulting in the center of mass where its linear velocity is null. The concept of the center of mass holds significant importance in the field of rotational dynamics due to its ability to streamline the analysis of complex object motion (Aravind & Heard, 2011).

To summarize, the moment of inertia is utilized to quantify the degree of resistance an object exhibits toward rotational motion. The strength of this resistance depends on how the object's mass is divided about its rotational axis. Various geometric shapes possess distinct mathematical formulas that can be employed to determine their respective moment of inertia. Torque and angular acceleration are the rotational counterparts to Newton's second law, with the total torque acting on an object being proportional to its angular acceleration. When an item is in a state of rotational equilibrium, its angular velocity remains constant, and it does not experience any angular acceleration when the net torque acting on it is zero. The center of mass streamlines the examination of rotational motion in intricate objects, offering a stationary reference point around which pure rotational motion occurs. A comprehensive grasp of the moment of inertia and rotational dynamics is imperative in order to analyze the characteristics of objects in rotation and to devise mechanical systems that incorporate rotational motion.

6.5. CONSERVATION OF ANGULAR MOMENTUM

Angular momentum (L) is a fundamental principle within the realm of rotational dynamics that serves to elucidate the rotational motion exhibited by an object or a collection of objects. The vector quantity in question is mathematically defined as the cross product between the moment of inertia (I) and the angular velocity (ω).

The equation $L = I \cdot \omega$ represents the relationship between the angular momentum vector (L), the angular velocity vector (ω), and the moment of inertia (I) about the axis of rotation.

Angular momentum possesses both magnitude and direction. The expression for the magnitude of angular momentum is

$L = I \cdot \omega,$

where L is the moment of inertia, I is the moment of angular momentum, and is the angular velocity. Also, the direction of angular momentum is not in the same plane as the vectors of moment of inertia and rotational velocity. To figure out the direction of angular motion, use the right-hand rule. This rule says that the thumb shows the direction of angular momentum when the fingers on the right hand curl in the direction of rotation (Figure 6.4) (Gao et al., 2006).

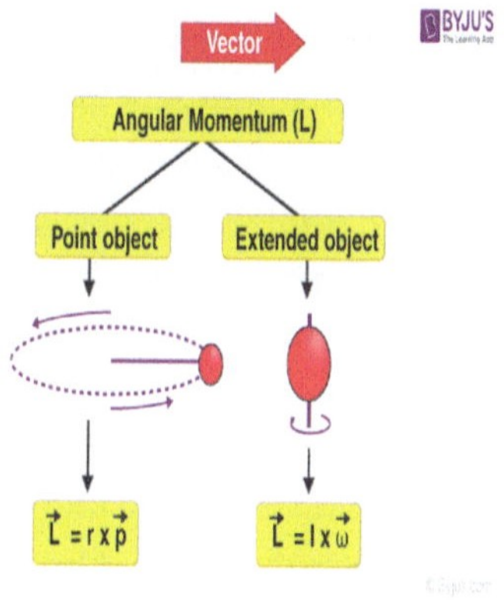

Figure 6.4. Schematic of the angular momentum (Source: Byju's Creative Commons License)

Angular momentum exhibits several significant properties:

1. The total angular momentum of a closed system does not change when no external torque is applied on the system, according to the principle of conservation of angular momentum. This principle is commonly referred to as the conservation of angular momentum and holds considerable implications for a range of physical phenomena.
2. The angular momentum is contingent upon the selection of the axis of rotation. The utilization of distinct axes can yield varying angular momenta for a given rotating entity.
3. Rotational symmetry is observed in objects that possess a preserved component of angular momentum along a particular axis. This characteristic facilitates the examination of rotational motion in such instances (Kirby, 2001).

6.5.1. Angular Momentum Conservation in Isolated Systems

The preservation of angular momentum is an essential concept in physics and is a pivotal factor in comprehending numerous natural phenomena. In the absence of external torques within an isolated system, the system's total angular momentum remains conserved over time. This principle is derived from using Newton's third law of motion and the rotational equivalent of Newton's second law.

The principle of conservation of angular momentum can be mathematically represented as follows:

The equation $\sum \tau ext=0$ denotes the summation of all external torques exerted on the system, resulting in a net torque of zero. When the sum of external torques is equal to zero ($\sum \tau_{ext}=0$), the net external torque acting on the system is zero, resulting in the conservation of the system's total angular momentum (Gerhard, 1981).

One of the fundamental outcomes of the conservation of angular momentum is that when an object or a system of objects experiences alterations in its rotational motion, its moment of inertia (I) and angular velocity (ω) adapt in order to uphold a consistent product of $I \cdot \omega$. As an illustration, when a rotating object undergoes a reduction in its moment of inertia, its angular velocity will correspondingly increase in order to uphold the principle of conservation of angular momentum (Kuzkin, 2015).

6.5.2. Spinning Objects, Rotating Skaters

Angular momentum is readily observable in a multitude of commonplace situations and carries intriguing ramifications for the dynamics of spinning entities and rotating individuals engaged in ice skating activities (Watanabe, 1992).

1. Spinning objects: The phenomenon of spinning objects, such as tops or gyroscopes, exhibiting precession around a vertical axis when tilted away from their vertical orientation can be attributed to the conservation of angular momentum.
2. Rotational skaters: Figure skaters demonstrate the principle of angular momentum conservation through their rotational movements. When figure skater retracts their arms towards their body, the moment of inertia of their body decreases. In order to preserve the conservation of angular momentum, the angular velocity of the system experiences an increase, resulting in an acceleration of its spin (Turro, 2011).

6.5.3. Gyroscopic Precession and Stability

The phenomenon of gyroscopic precession is a captivating occurrence that emerges as a consequence of the principle of conservation of angular momentum. The phenomenon arises when a rotating object undergoes a torque that is exerted in a direction perpendicular to its rotational axis. Consequently, the angular momentum vector of the object experiences a progressive rotation or precession in relation to the direction of the exerted torque.

Remember

Gyroscopic precession is a fascinating phenomenon observed in spinning objects, which leads to stability and control in various devices like bicycles and spacecraft.

The phenomenon above is employed in gyroscopes, instruments that harness gyroscopic precession for diverse purposes. Gyroscopes use extensively in navigation systems, spacecraft stabilization, and various other engineering applications. The capacity to sustain a consistent spatial orientation renders them highly valuable instruments in the field of aerospace engineering (Figure 6.5).

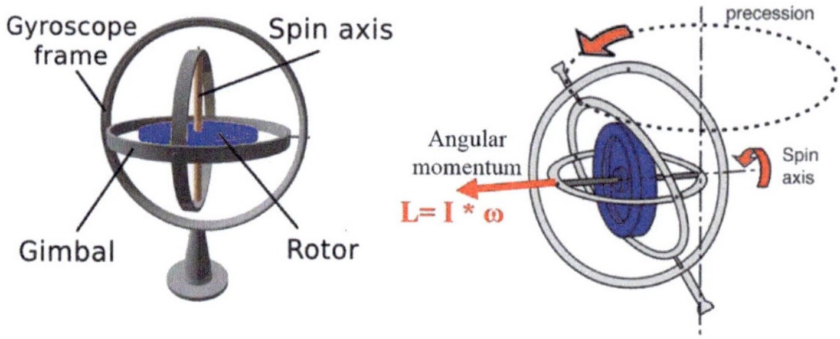

Figure 6.5. Gyroscope components and gyroscopic precession (Source: Felix Hamza Lup, Creative Commons License)

Gyroscopic stability is observed in a range of objects, including bicycles and motorcycles. The rotational motion of the wheels in these vehicles serves as gyroscopes, contributing to their stability by resisting tilting and preserving their alignment in the direction of travel (Fang et al., 2013).

In summary, angular momentum is a fundamental principle within the realm of rotational dynamics, representing the rotational motion exhibited by an individual object or a collection of objects. The conservation of angular momentum in isolated systems has significant implications for comprehending the dynamics of spinning objects, rotating skaters' motion, and gyroscope stability. The principle of angular momentum conservation is an important concept in elucidating various natural events and engineering applications. The influence of angular momentum is observable in diverse domains of physics and engineering, encompassing celestial motion, spinning objects, and gyroscopic devices (Jin et al., 2016).

6.6. ROTATIONAL WORK, ENERGY, AND POWER

In the context of rotational motion, the principles of work, energy, and power exhibit similarities to their corresponding principles in translational motion. Applying torque to a rotating object induces work, leading to a modification in its rotational energy. The quantification of an object's rotational energy is determined by its rotational kinetic energy, which is contingent upon both its moment of inertia and angular velocity.

According to the definition of work, rotational work is the transfer of energy that happens when a force is given over a distance. Work in translational motion is given by the equation $W = Fdcos()$, where W is the amount of work done, F is the amount of force applied, d is the distance the object moves, and is the angle between the force vector and the distance vector. In the context of rotational motion, it is possible to establish a corresponding equation for rotational work by incorporating the concept of torque (τ). This equation can be expressed as follows: $W = \tau \cdot \theta$. The rotational work done, denoted as W, is determined by the torque applied, represented by τ, and the angular displacement of the rotating object, denoted as θ (Mehigan et al., 2020).Rotational kinetic energy refers to the energy attributed to an object's rotational motion. The concept resembles the linear kinetic energy exhibited by an object in motion. The expression for the rotational kinetic energy of an object can be determined by multiplying half of the moment of inertia (I) by the square of the angular velocity (ω).The relationship expressed by this equation demonstrates that the rotational kinetic energy is contingent upon both the moment of inertia of the object and its angular velocity (Cao et al., 2022).

6.6.1. Kinetic Energy of Rotating Bodies

The comparison between rotational kinetic energy and linear kinetic energy of a moving object can be made. The linear kinetic energy, denoted as KE_{lin}, is mathematically expressed as $1/2mv^2$, where m represents the mass of the object and v represents its linear velocity.

In order to gain a deeper comprehension of the correlation between rotational and linear kinetic energy, let us examine the dynamics of a point mass (m) traversing a circular path with a radius (r) possessing both angular velocity (ω) and linear velocity (v).

The linear velocity can be mathematically expressed as the product of the radius and the angular velocity, denoted as $v = r \cdot \omega$. The equation for the moment of inertia of a point mass rotating about an axis perpendicular to its path is given by $I = m \cdot r^2$.

By substituting these relationships into the formula for rotational kinetic energy, we can obtain a more comprehensive understanding of the concept.:

$$KE_{rot}=I\omega^2=(m\cdot r^2)\omega^2=m(r\cdot\omega)^2=mv^2=KE_{lin}$$

This finding demonstrates that the rotational kinetic energy is equivalent to the linear kinetic energy in circular motion in the case of a point mass in a circular motion. Put differently, the aggregate kinetic energy of a rotating entity can be conceptualized as the summation of the linear kinetic energies possessed by each of its constituent point masses (Ferronsky et al., 1996).

6.6.2. Rotational Power and Its Application in Mechanical Systems

Power (P) refers to the measure of the rate at which work is performed or the rate at which energy is transferred or converted. In the context of rotational motion, the concept of power resembles its equivalent in translational motion.

Rotational power refers to the power linked to an object's rotational motion. It can be determined by multiplying the applied torque (τ) with the object's angular velocity (ω). Mathematically, the equation representing rotational power is given as $P=\tau\cdot\omega$, where P represents the rotational power, τ represents the applied torque, and ω represents the angular velocity of the rotating object.

The Concept of Mechanical Advantage in Mechanical Systems: Power plays a pivotal role in mechanical systems, as it signifies the capacity to execute work or transfer energy. In various mechanical systems, such as motors and engines, the utilization of power is employed to generate mechanical work. The quantity that represents the relationship between the power output and power input is referred to as mechanical advantage, which plays a crucial role in determining the effectiveness of the mechanical system (Chen et al., 2020).

A higher mechanical advantage implies that the system has the capability to convert a greater proportion of the input power into productive work, resulting in enhanced efficiency. Engineers frequently engage in the optimization of mechanical systems in order to attain a heightened mechanical advantage and enhance overall performance (Nikravesh & Chung, 1982).

6.6.3. Conservation of Mechanical Energy in Rotational Motion

The principle of work energy in translational motion elucidates the conservation of mechanical energy. According to this principle, the net-work exerted on an object is equivalent to its alteration in kinetic energy (KE_{lin}):

$$\Delta KE_{lin}=W_{net}$$

In rotational motion, a similar principle applies to the conservation of mechanical energy, but it involves the change in rotational kinetic energy (KE_{rot}) instead:

$$\Delta KE_{rot}=W_{rot}$$

where ΔKE_{rot} is the change in rotational kinetic energy, and W_{rot} is the total rotational work done on the object.

For an isolated system where no external torques act, the conservation of mechanical energy in rotational motion can be expressed as (Puttharugsa & Pimanpang, 2021):

$$\Delta KE_{rot}=0$$

The equation posits that in the absence of external torques exerted on the rotating system, the overall rotational kinetic energy remains unchanging, thereby conserving the system's mechanical energy.

The principle of mechanical energy conservation holds considerable implications for comprehending the dynamics of rotating entities, such as spinning tops or celestial bodies in their orbital paths. This approach facilitates the anticipation of the movement of said objects, circumventing the need to account for intricate forces, thereby rendering it an influential instrument in the examination of intricate rotational systems (Pierratos & Polatoglou, 2017).

6.7. TORQUE AND EQUILIBRIUM

The concept of torque holds significant importance in the field of rotational dynamics as it pertains to the efficacy of a force in generating rotational motion around an axis (Figure 6.6). The rotational analog of force in translational motion can be observed. The quantity known as torque (τ) can be mathematically expressed as the vector cross product of the force (F) exerted on an object and the lever arm (r) of said force relative to the axis of rotation:

$$\tau=r\times F$$

In the given context, τ represents the torque vector, **r** denotes the vector extending from the axis of rotation to the location where the force is applied, and **F** represents the force vector.

The characteristics of torque are discussed below.

Torque is classified as a vector quantity characterized by its direction, which is orthogonal to both the lever arm vector (**r**) and the force vector(**F**). The direction of torque adheres to the right-hand rule. When the fingers of the right hand align with the lever arm vector and rotate in the direction of the force vector, the thumb indicates the direction of the torque vector (Parfrey et al., 2016).

Figure 6.6. Types of equilibrium (Source: Kofa Study, Creative Commons License)

The magnitude of torque can be expressed as the product of the magnitudes of the lever arm vector and the force vector, multiplied by the sine of the angle between them, i.e., $|\tau|=|\mathbf{r}|\cdot|\mathbf{F}|\cdot\sin(\theta)$. The maximum torque is achieved when the angle between the applied force and the lever arm is 90 degrees, resulting in a sine value of 1 (Wereszczynski & Andricioaei, 2006).

6.7.1. Conditions for Equilibrium in Rotational Systems

Equilibrium in rotational systems pertains to a condition in which an object remains stationary or rotates at a consistent angular velocity. Within the current state, the object experiences an absence of net torque, leading to a consequential absence of angular acceleration.

In order to achieve equilibrium in rotational systems, it is necessary to satisfy two conditions: The net torque (τ_{net}) is equal to zero. In the context of rotational equilibrium, the algebraic sum of all torques exerted on the object must be equal to zero.

In a mathematical context, this can be represented as the summation of τ from 0. The symbol $\sum\tau$ denotes the summation of all torques exerted on the object (Yang et al., 2021).

6.7.2. Static Equilibrium: Analyzing Balanced Torques

The concept of static equilibrium pertains to the condition of an object wherein it remains motionless and exhibits no linear or angular acceleration. In a state of static equilibrium, the object experiences a condition where the forces and torques exerted upon it are in equilibrium, leading to the absence of any resultant force or torque. Static equilibrium is classified into two distinct types:

1. Translational static equilibrium: In the context of mechanics, translational static equilibrium refers to the state in which an object remains motionless, with the total force exerted on it being equal to zero. This implies that the resultant of all the forces in both the horizontal and vertical components is equal to zero.
2. Rotational static equilibrium: Rotational static equilibrium is the state in which an object remains stationary, with the condition that the total torque exerted on it is equal to zero. It is necessary for the total sum of torques acting about any given axis of rotation to be equal to zero (Lin et al., 2009).

Examining Equilibrium through the Analysis of Balanced Torques:

In the state of rotational equilibrium, it is necessary for the torques exerted on an object to be in a state of balance. This implies that irrespective of the selected axis of rotation, the magnitudes of the torques acting in the clockwise direction must be equivalent to the magnitudes of the torques acting in the counterclockwise direction. In the event that the torques acting on an object are not balanced, the object will undergo angular acceleration and will not be in a state of equilibrium. In order to examine the state of balanced torques in a state of equilibrium, it is necessary to consider both the lever arms and the magnitudes of the forces exerted on the object. The prudent selection of the rotation axis is imperative to streamline the analysis process and ascertain the equilibrium of torques (de Almeida Lins et al., 2013).

6.7.3. Applications of Equilibrium Principles in Engineering

The principles of equilibrium, particularly static equilibrium, find extensive applications in engineering and design. Engineers employ these fundamental principles in order to devise structures and mechanical systems that maintain stability and equilibrium across diverse circumstances. Several common applications are typically utilized.

1. Building and Structural Design: The field of building and structural design involves the application of equilibrium principles by architects and civil engineers. These professionals utilize these principles to ensure the stability and safety of various structures, such as buildings and bridges. The structures are subjected to a thorough analysis of the forces and torques in order to mitigate the risk of structural errors and collapse (Denbigh, 1981).

2. Mechanical Systems: Mechanical systems encompass the application of equilibrium principles within the field of mechanical engineering. These principles are utilized to devise machines and processes that effectively uphold balance and stability throughout their operational processes. Illustrative instances encompass cranes, automotive suspensions, and robotic arms.
3. Bridge Design: Bridge design involves the application of equilibrium principles by engineers in order to create structures capable of withstanding applied loads without experiencing significant deformation or failure. The bridge components are subjected to a thorough analysis of the forces exerted upon them, with particular attention given to ensuring the equilibrium of torques (Sharma et al., 1999).
4. Weight distribution: Weight distribution is a critical aspect in industries such as aviation and automotive, where equilibrium principles are employed to achieve appropriate weight distribution in vehicles and aircraft. Ensuring a balanced weight distribution is of utmost importance in order to achieve stability and ensure the safe operation of the system.

Did You Know?

The principles of equilibrium in rotational systems find applications in engineering designs, such as creating stable structures like bridges and buildings.

5. Architecture: In the field of architecture, the utilization of equilibrium principles is employed as a means to create attractive structures and artwork that uphold stability and balance.

6.7.4. Rotational Motion of Rigid Bodies

Rigid body rotations pertain to the rotational dynamics exhibited by elongated objects, such as a wheel in rotation, a disk undergoing circular motion, or a pendulum swinging back and forth. In contrast to a point mass, which undergoes uncomplicated rotational motion around a stationary axis, rigid bodies exhibit intricate rotational dynamics as a result of their elongated form and mass distribution.

In the context of rigid body rotations, it can be observed that the constituent particles of an object undergo rotational motion around a shared axis of rotation. The axis in question is commonly referred to as the instantaneous axis of rotation or the axis of rotation at a specific moment. Various points on the rigid body exhibit distinct linear velocities while collectively possessing a uniform angular velocity (ω) concerning the axis (Argyris, 1954).

A rigid body's angular velocity (ω) is formally defined as the derivative of the angular displacement concerning time, indicating the rate at which the object undergoes rotational motion. This physical quantity is conventionally expressed in radians per second (rad/s) units. The equation $v = r \cdot \omega$ represents the mathematical relationship between the linear velocity (v) and the angular velocity (ω) of a point located at a distance r from the axis of rotation (Velasco et al., 2014).

Activity 1.1

A merry-go-round in a playground has children of different ages and sizes riding on it. The merry-go-round is initially at rest, and then one child gives it a strong push to set it in motion. As the merry-go-round starts spinning, the children move towards the outer edge. Explain why the children move towards the outer edge of the merry-go-round as it spins.

SUMMARY

This chapter introduces the concept of rotational motion and explores its comparison with translational motion. It delves into rotational kinematics, discussing angular and linear quantities, kinematic equations, and angular acceleration in circular motion. The chapter also covers moment of inertia and rotational dynamics, including torque, angular acceleration, rotational equilibrium, and center of mass. Conservation of angular momentum in isolated systems is explained, along with gyroscopic precession and stability. Rotational work, energy, and power are explored, focusing on kinetic energy of rotating bodies and conservation of mechanical energy in rotational motion. The chapter concludes with discussions on torque and equilibrium, conditions for equilibrium in rotational systems, and applications of equilibrium principles in engineering.

REVIEW QUESTIONS

1. What are the differences between rotational motion and translational motion, and how are they related?

2. How do you calculate the moment of inertia for different shapes, and how does it influence rotational dynamics?
3. Explain the concept of angular momentum conservation in isolated systems, and provide real-world examples.
4. What is gyroscopic precession, and what role does it play in the stability of rotating objects?
5. How does rotational work, energy, and power relate to kinetic energy and mechanical energy conservation?
6. Describe the conditions for equilibrium in rotational systems, and how are these principles applied in engineering scenarios?

REFERENCES

1. Aravind, V. R., & Heard, J. W. (2011). Mass balance with a meterstick: Teaching rotational equilibrium. *Latin-American Journal of Physics Education*, *5*(4), 686–688.
2. Argyris, J. H. (1954). Energy Theorems and Structural Analysis: A Generalized Discourse with Applications on Energy Principles of Structural Analysis Including the Effects of Temperature and Non-Linear Stress-Strain Relations. *Aircraft Engineering and Aerospace Technology*, *26*(10), 347–356.
3. Bills, B. G., & James, T. S. (1999). Moments of inertia and rotational stability of Mars: Lithospheric support of subhydrostatic rotational flattening. *Journal of Geophysical Research: Planets*, *104*(E4), 9081–9096.
4. Bishop, D. M. (1990). Molecular vibrational and rotational motion in static and dynamic electric fields. *Reviews of Modern Physics*, *62*(2), 343.
5. Cao, B., Wang, P., Rui, P., Wei, X., Wang, Z., Yang, Y., ... & Wang, Z. L. (2022). Broadband and Output-Controllable Triboelectric Nanogenerator Enabled by Coupling Swing-Rotation Switching Mechanism with Potential Energy Storage/Release Strategy for Low-Frequency Mechanical Energy Harvesting. *Advanced Energy Materials*, *12*(46), 2202627.
6. Chen, C., Ma, T., Jin, H., Wu, Y., Hou, Z., & Li, F. (2020). Torque and rotational speed sensor based on resistance and capacitive grating for rotational shaft of mechanical systems. *Mechanical Systems and Signal Processing*, *142*, 106737.
7. Cochard, A., Igel, H., Schuberth, B., Suryanto, W., Velikoseltsev, A., Schreiber, U., ... & Vollmer, D. (2006). Rotational motions in seismology: theory, observation, simulation. *Earthquake Source Asymmetry, Structural Media and Rotation Effects*, 391–411.
8. Damavandi, M., Barbier, F., Leboucher, J., Farahpour, N., & Allard, P. (2009). Effect of the calculation methods on body moment of inertia estimations in individuals of different morphology. *Medical Engineering & Physics*, *31*(7), 880–886.
9. de Almeida Lins, C. A., Neto, F. L., de Amorim, A. B. C., de Brito Macedo, L., & Brasileiro, J. S. (2013). Kinesio Taping® does not alter neuromuscular performance of femoral quadriceps or lower limb

function in healthy subjects: Randomized, blind, controlled, clinical trial. *Manual Therapy, 18*(1), 41–45.

10. Denbigh, K. G. (1981). *The Principles of Chemical Equilibrium: With Applications in Chemistry and Chemical Engineering*. Cambridge University Press.
11. Fang, J., Ren, Y., & Fan, Y. (2013). Nutation and precession stability criterion of magnetically suspended rigid rotors with gyroscopic effects based on positive and negative frequency characteristics. *IEEE Transactions on Industrial Electronics, 61*(4), 2003–2014.
12. Ferronsky, V. I., Denisik, S. A., & Ferronsky, S. V. (1996). Virial oscillations of celestial bodies: V. The structure of the potential and kinetic energies of a celestial body as a record of its creation history. *Celestial Mechanics and Dynamical Astronomy, 64*, 167–183.
13. Frobin, W., Leivseth, G., Biggemann, M., & Brinckmann, P. (2002). Sagittal plane segmental motion of the cervical spine. A new precision measurement protocol and normal motion data of healthy adults. *Clinical Biomechanics, 17*(1), 21–31.
14. Gao, F., Latash, M. L., & Zatsiorsky, V. M. (2006). Maintaining rotational equilibrium during object manipulation: linear behavior of a highly non-linear system. *Experimental Brain Research, 169*, 519–531.
15. Gerhard, O. E. (1981). N-body simulations of disc–halo galaxies: isolated systems, tidal interactions and merging. *Monthly Notices of the Royal Astronomical Society, 197*(1), 179–208.
16. Hadziioannou, C., Gaebler, P., Schreiber, U., Wassermann, J., & Igel, H. (2012). Examining ambient noise using colocated measurements of rotational and translational motion. *Journal of Seismology, 16*, 787–796.
17. Hajiaghamemar, M., Seidi, M., & Margulies, S. S. (2020). Head rotational kinematics, tissue deformations, and their relationships to the acute traumatic axonal injury. *Journal of Biomechanical Engineering, 142*(3), 031006.
18. Hasan, D. S., & Hamarash, I. (2017). Calculation of angular velocity, angular acceleration and torque of two common point rigid bodies using IMU. *Journal of Applied and Physical Sciences, 3*(2), 58–64.
19. Jin, H., Wang, T., Yu, F., Zhu, Y., Zhao, J., & Lee, J. (2016). Unicycle robot stabilized by the effect of gyroscopic precession and its control realization based on centrifugal force compensation. *IEEE/ASME Transactions on Mechatronics, 21*(6), 2737–2745.

20. Kirby, K. A. (2001). Subtalar joint axis location and rotational equilibrium theory of foot function. *Journal of the American Podiatric Medical Association*, *91*(9), 465–487.

21. Kolyada, Y. B., Korolev, Y. S., Krushnyak, N. T., Razgulin, V. G., & Yanushkin, V. N. (2010). Prospects and problems of automation, improving response speed, accuracy, and reliability of devices for measuring linear and angular quantities. *Measurement Techniques*, *53*(8), 858.

22. Kuzkin, V. A. (2015). On angular momentum balance for particle systems with periodic boundary conditions. *ZAMM-Journal of Applied Mathematics and Mechanics/Zeitschrift für Angewandte Mathematik und Mechanik*, *95*(11), 1290–1295.

23. Leclercq, G., Lefèvre, P., & Blohm, G. (2013). 3D kinematics using dual quaternions: theory and applications in neuroscience. *Frontiers in Behavioral Neuroscience*, *7*, 7.

24. Lin, W. H., Liu, Y. F., Hsieh, C. C. C., & Lee, A. J. (2009). Ankle eversion to inversion strength ratio and static balance control in the dominant and non-dominant limbs of young adults. *Journal of Science and Medicine in Sport*, *12*(1), 42–49.

25. Mashood, K. K., & Singh, V. A. (2012). An inventory on rotational kinematics of a particle: unravelling misconceptions and pitfalls in reasoning. *European Journal of Physics*, *33*(5), 1301.

26. Mehigan, L., Al Kez, D., Collins, S., Foley, A., Ó'Gallachóir, B., & Deane, P. (2020). Renewables in the European power system and the impact on system rotational inertia. *Energy*, *203*, 117776.

27. Mulhayatiah, D., Suhendi, H. Y., Zakwandi, R., Dirgantara, Y., & Ramdani, M. A. (2018). Moment of inertia: development of rotational dynamics KIT for physics students. In *IOP Conference Series: Materials Science and Engineering* (Vol. 434, No. 1, p. 012014). IOP Publishing.

28. Nikravesh, P. E., & Chung, I. S. (1982). Application of Euler parameters to the dynamic analysis of three-dimensional constrained mechanical systems.

29. Nutsathaporn, C., Chomkokard, S., Wongkokua, W., Ruengittinun, S., & Jinuntuya, N. (2023). Monte Carlo study of uncertainty propagation in Euler rotational kinematic equation. In *Journal of Physics: Conference Series* (Vol. 2431, No. 1, p. 012036). IOP Publishing.

30. Parfrey, K., Spitkovsky, A., & Beloborodov, A. M. (2016). Torque enhancement, spin equilibrium, and jet power from disk-induced opening of pulsar magnetic fields. *The Astrophysical Journal*, *822*(1), 33.
31. Pierratos, T., & Polatoglou, H. M. (2017). Study of the conservation of mechanical energy in the motion of a pendulum using a smartphone. *Physics Education*, *53*(1), 015021.
32. Puttharugsa, C., & Pimanpang, S. (2021). Analyzing the rotational motion of a rectangular board via smartphone sensors: a conservation-of-mechanical-energy approach. *Physics Education*, *56*(2), 023005.
33. Rani, M. U. (2021). Calculation of Moment of Inertia for various Geometrical cross section using C Programming. *International Research Journal of Engineering and Technology (IRJET)*, *8*(11).
34. Sharma, R., Singhal, D., Ghosh, R., & Dwivedi, A. (1999). Potential applications of artificial neural networks to thermodynamics: vapor–liquid equilibrium predictions. *Computers & Chemical Engineering*, *23*(3), 385–390.
35. Turro, N. J. (2011). Fun with photons, reactive intermediates, and friends. Skating on the edge of the paradigms of physical organic chemistry, organic supramolecular photochemistry, and spin chemistry. *The Journal of Organic Chemistry*, *76*(24), 9863–9890.
36. Velasco, J. A., Lopez, L., Cabrera, S., Boutonnet, M., & Järås, S. (2014). Synthesis gas production for GTL applications: Thermodynamic equilibrium approach and potential for carbon formation in a catalytic partial oxidation pre-reformer. *Journal of Natural Gas Science and Engineering*, *20*, 175–183.
37. Watanabe, J. I. (1992). Ice-skater model for the nucleus of Comet Levy 1990c-Spin-up by a shrinking nucleus. *PASJ: Publications of the Astronomical Society of Japan (ISSN 0004–6264), vol. 44,* no. 2, 1992, p. 163–166., *44*, 163–166.
38. Wereszczynski, J., & Andricioaei, I. (2006). On structural transitions, thermodynamic equilibrium, and the phase diagram of DNA and RNA duplexes under torque and tension. *Proceedings of the National Academy of Sciences*, *103*(44), 16200–16205.
39. Yang, Y., Huang, L., Xiang, J., Bao, H., & Li, H. (2021). Design of multi-wing 3D chaotic systems with only stable equilibria or no equilibrium point using rotation symmetry. *AEU-International Journal of Electronics and Communications*, *135*, 153710.

40. Zhang, H., Fang, H., Fang, Y., & Jiang, B. (2018). Workspace analysis of a hybrid kinematic machine tool with high rotational applications. *Mathematical Problems in Engineering, 2018*.

CHAPTER 7

Oscillation Simple Harmonic Motion

Contents

7.1. INTRODUCTION

Periodic or harmonic motion refers to any movement that repeats itself at regular intervals. The period of such motion corresponds to the interval after that the motion is replicated, resulting in the same position and velocity of the moving object. Several instances of periodic motion can be observed, as illustrated in Figure 7.1.

1. The orbital dynamics of celestial bodies revolve around the central star.
2. The reciprocating motion of a piston within a cylinder is commonly observed in internal combustion engines utilized in automotive applications.
3. The study concerns the movement of a spherical object within a concave container (Festiana et al., 2019).

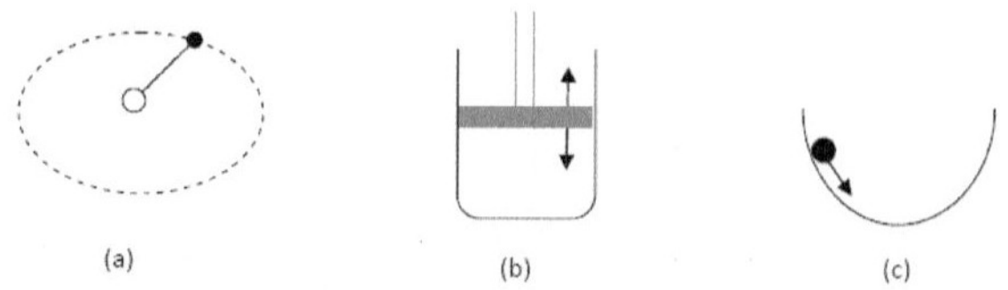

Figure 7.1. Several examples of periodic motion can be identified. (a) The motion of celestial bodies such as the earth revolving around the sun or the moon orbiting the earth can be considered periodic. (b) The reciprocating motion of a piston within a cylinder (Source: S. C. Garg, Creative Commons License)

In the event of periodic motion, when a body undergoes repeated back-and-forth movement around a fixed position known as equilibrium or mean position, the motion is classified as oscillatory or vibratory. Examples of periodic motion include the earth's rotation about the sun and the movement of clock hands. However, it is important to note that these examples are not characterized by oscillatory behavior. Examples of oscillatory motion include the movement of a piston in a vehicle's engine, the motion of a ball within a bowl, the oscillation of a needle in a sewing machine, and the swinging motion of the bob in a pendulum clock (Earley, 2023).

Simple harmonic motion (SHM) refers to the motion of an oscillating body when the forces acting on it are directly proportional to its movement from the mean position, and the restoring force always points towards the mean position. Therefore, it is evident that simple harmonic motion (SHM) can be regarded as a specific instance of oscillatory and vibratory motion. In this unit, a comprehensive examination of Simple Harmonic Motion (SHM) will be conducted. Several instances of simple harmonic motion can be observed, as depicted in Figure 7.2 (Getachew, 2018).

The two examples of systems that exhibit periodic motion are:

1. the movement of a simple pendulum; and
2. the vibration of a tuning fork.

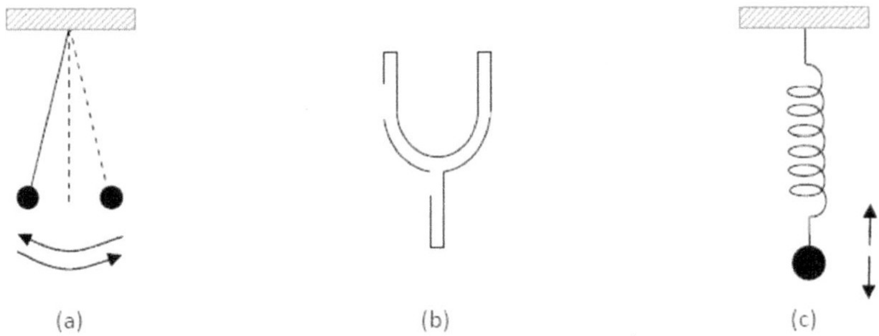

Figure 7.2. Some examples of SHM: (a) A simple pendulum; (b)a vibrating tuning fork; (c) an oscillating spring-mass system (Source: S.R. Jha, Creative Commons License)

7.2. OSCILLATORY MOTION

An oscillating system exhibits repetitive to-and-fro motion. Oscillations can exhibit a high degree of complexity, as observed in phenomena such as the vibrations of a piano string, the seismic activity of the Earth through an earthquake, or the rhythmic contractions of the heart. There exist oscillations that are imperceptible to human senses, such as the oscillations of air molecules responsible for transmitting sound, the oscillations of atoms in solids that convey temperature sensations, and the oscillations of electrons in radio and TV transmitter antennas. It is not an overstatement to assert that oscillations pervade our surroundings consistently, as oscillatory phenomena extend beyond physical entities like musical instruments. Visible light, microwaves, radio waves, and X-rays are also manifestations of oscillatory processes. Therefore, the investigation of oscillations is imperative for comprehending diverse systems, encompassing mechanical, acoustical, electrical, and atomic domains (Womersley, 1955).

The occurrence of oscillatory motion in a physical system can be attributed to two fundamental properties, namely inertia, and elasticity. The subsequent sections will introduce two physical systems that serve as illustrative examples. Examining these elementary systems will facilitate our comprehension of the dynamics exhibited by more intricate oscillatory systems (Evans & Michard, 1962).

7.2.1. Simple Pendulum

Do you recall conducting an experiment involving a simple pendulum during your senior secondary class in the physics laboratory, wherein you calculated the variation in the time period in relation to the length of the string?

Tip

The period of a simple pendulum depends on its length, and it remains constant as long as the amplitude is small.

A basic pendulum is comprised of a massive point mass that is hanging from a stationary support by means of a string that is both weightless and incapable of being stretched. It is imperative to comprehend that a simple pendulum is a theoretical construct. In practical applications, the implementation of a simple pendulum involves the suspension of a small metallic sphere using a thread that is affixed to a stationary support structure, such as a stand. Figure 7.3 depicts a basic pendulum system, wherein a mass m is affixed to a stationary support P by means of a slender string with negligible mass and a length l. In its natural state, the bob is situated at position PO, characterized by an angle = 0, commonly referred to as the mean and equilibrium position. Starting from the equilibrium position, the pendulum is displaced towards one extreme, denoted as A, while ensuring that the angle remains within a small range. By doing so, the bob acquires a certain amount of finite potential energy (Vogt & Kuhn, 2012). Upon the release of the bob from point A, it initiates a downward motion towards the equilibrium position denoted as O. Consequently, the potential energy of the bob diminishes as it nears point O. Since the potential energy diminishes, the bob acquires kinetic energy. At the equilibrium position, the kinetic energy of the bob reaches its maximum value while its potential energy reaches its minimum value. Further. As a result of acquiring kinetic energy, the pendulum bob does not come to a halt at the equilibrium position; instead, it surpasses the equilibrium position and proceeds to reach the opposite extreme point B. At position B, the potential energy of Bob reaches its maximum value, while the kinetic energy becomes zero due to the momentary cessation of his velocity. Following a brief period of respite, the bob proceeds to retrace its trajectory from point B to point O and subsequently to point A. Therefore, it can be observed that the bob undergoes oscillatory motion along a circular trajectory, with the point of suspension P serving as the center (Lima & Arun, 2006).

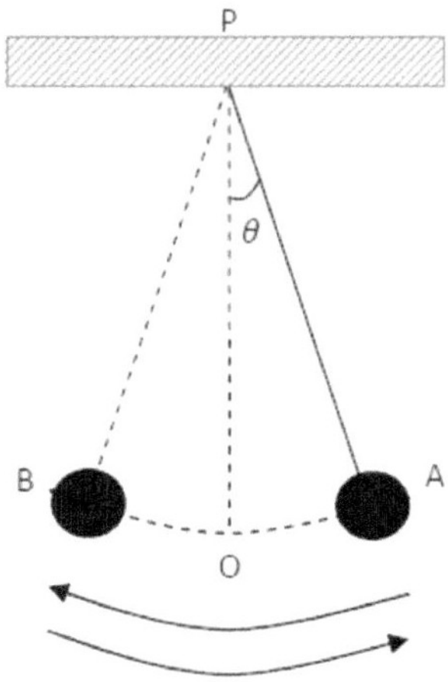

Figure 7.3. A simple pendulum (Source: Pant, Creative Commons License)

In an ideal scenario characterized by the absence of air resistance, the presence of friction does not exert any influence on the oscillatory motion. In this particular scenario, it can be postulated that the pendulum has the potential to oscillate indefinitely. Every full oscillation of its motion carries it from one equilibrium position to the opposite side and subsequently returns it to its original position (Brodie et al., 2008).

7.2.2. Spring-Mass System

It is imperative to possess knowledge of the spring pendulum, in addition to the simple pendulum.

The spring mass system, also known as the spring pendulum, is comprised of a weightless spring with a constant k. One end of the spring is securely fixed to a wall, while the other end is connected to a body of mass m. The body is able to move either horizontally or vertically, according to the specific system. When a spring is under tension, it has the capability to oscillate in a vertical direction. In the scenario of a horizontal spring-mass system, the object is able to undergo motion on a horizontal surface devoid of friction, as depicted in Figure 7.4. When a spring is extended, the inherent property of elasticity within the spring applies a force that acts to reinstate the mass to its equilibrium position. Upon reaching the mean position, the mass has acquired a certain velocity. Consequently, the mass persists in its motion along the same trajectory, ultimately causing the spring to undergo compression until it attains the opposite extreme position. The compressed spring exerts a restorative force on the mass, causing it to move in the direction opposite to its displacement and return to its equilibrium position. Therefore, every oscillatory cycle involves the displacement of mass m from a particular extreme position to the opposing extreme position relative to the mean position (Gürgöze, 1996).

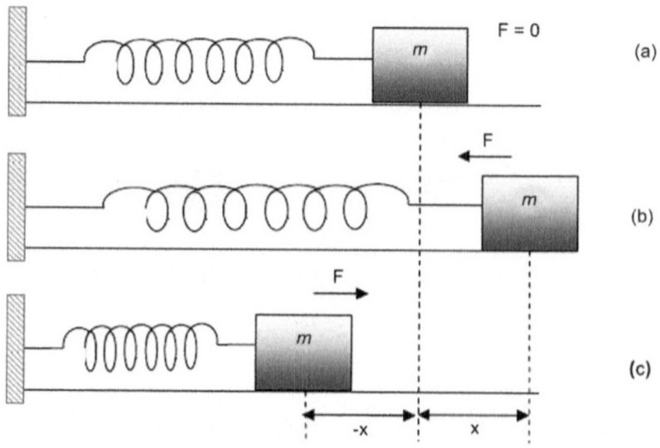

Figure 7.4. (a) Normal, (b)stretched, (c) compressed configurations of a horizontal spring-mass system (Source: Gairola, Creative Commons License)

In an ideal scenario, characterized by the absence of air resistance and a frictionless horizontal surface, the spring-mass system would exhibit perpetual oscillation.

7.3. SIMPLE HARMONIC MOTION

In the preceding section, two instances of oscillatory motion were examined. In the present discourse, we shall employ the spring-mass system as a means to comprehend the phenomenon of simple harmonic motion (SHM). What is the concept of Simple Harmonic Motion (SHM)? First and foremost, let us proceed to address this inquiry (Triana & Fajardo, 2011).

7.3.1. Definition of SHM

Simple Harmonic Motion (SHM) can be described and characterized using various definitions and explanations.

1. If the force exerted on the oscillating object consistently opposes the object's displacement from its equilibrium or mean position, as well as the force's magnitude is directly proportional to the displacement's magnitude, the object is said to be undergoing Simple Harmonic Motion (SHM).
2. If the graph of dislocation versus time for the oscillating object follows a sinusoidal pattern, it can be classified as undergoing Simple Harmonic Motion (SHM). This constitutes an alternative explication of Simple Harmonic Motion (SHM) (Adolphus et al., 2013).
3. If the potential energy of an oscillating body is directly proportional to the square of its displacement relative to the mean position, it is classified as performing Simple Harmonic Motion (SHM). This represents an additional interpretation of Simple Harmonic Motion (SHM) (Musik, 2017).

Let us now direct our attention to the initial definition. The spring-mass system depicted in Figure 7.4 (a) is in a state of static equilibrium, where the spring is at its natural length without any elongation or compression, and there is an absence of external forces acting on the mass. When a force is applied to displace the body to the right by a small distance x, the spring exhibits characteristics similar to those of an elastic system experiencing stress. One may recollect Young's modulus experiment conducted in a school setting. It is well-established that Young's modulus Y of a wire material can be determined by the relationship between the force magnitude F applied to stretch a wire of length L through a distance x (Lazos et al., 2022).

$$Y = \frac{F/\alpha}{x/L}$$

By manipulating the terms in the aforementioned equation, we can readily obtain the subsequent form

$$F = \left(\frac{Y\alpha}{L}\right)x$$

The concept of elasticity refers to the inherent characteristic of a body to resist alterations in its dimensions or configuration, thereby enabling the body to restore its initial state once the external force causing deformation is eliminated, provided that the force remains within a specified threshold. In alternative terms, it can be stated that when a body is in a deformed state, it generates a force that acts to restore its original shape. This restoring force, according to the Newton's third law of motion, is equal in magnitude but opposing in direction to the force that caused the deformation. It can be inferred that the restoring force is directly proportional to the lengthening and is oriented towards the equilibrium position, which is the state of rest in the absence of any restoring force. In the context of the spring-mass system, it can be observed that Hooke's Law postulates a direct relationship between the restoring force and the movement of the spring, regardless of whether the spring is stretched or compressed. In the present scenario, the spring exerts a restoring force on the body that is oriented towards the left, as depicted in Figure 7.4 (b). and can be expressed by the following equation (Reyes, 2019):

$$F=-kx$$

The motion is considered simple harmonic due to the proportional relationship between the restoring force, F, and the displacement, as well as the opposite signs of these quantities. The variable k is commonly referred to as the spring constant and stiffness constant. The SI unit for the variable k is expressed as Newton per meter (Nm^{-1}) (Wang et al., 2017).

7.3.2. Basic Characteristics of SHM

Having acquired an understanding of Simple Harmonic Motion (SHM), it is now appropriate to establish a clear definition of several fundamental attributes associated with SHM.

What thoughts are evoked in your mind? The primary significant attribute in simple harmonic motion (SHM) is the initial displacement, which serves as the impetus for the ensuing oscillatory behavior. The amplitude (A) of oscillations refers to the magnitude of the original displacement, that is also equivalent to the maximum displacement. As previously stated, the energy of the system undergoing Simple Harmonic Motion (SHM) oscillates between kinetic as well as potential manifestations. At the limits of the oscillatory motion, the kinetic energy reaches a minimum value of zero due to the velocity being zero, while the potential energy attains its maximum value (Saputri et al., 2021).

Remember

In simple harmonic motion (SHM), the acceleration of the particle is directly proportional to its displacement but in the opposite direction.

Another notable attribute of simple harmonic motion (SHM) is the time period (T), which represents the duration required for a single full cycle of oscillation. The minimum duration required for an oscillating object to transition from a specific position as well as velocity to its original position as well as velocity is referred to as the least time taken.

Typically, in order to facilitate ease of measurement, the time period is assessed either from the mean position or the extreme ends.

Instead of referring to a specific time period, it is common to utilize the frequency (v) as a means of characterizing simple harmonic motion (SHM) on numerous occasions. Frequency refers to the quantity of full oscillations performed within a unit of time, typically measured in cycles per second. It is the reciprocal of the time period (Andrews & Romero, 2009).

v=

The unit of measurement is denoted as cycles per second, symbolized as s^{-1} or hertz (Hz). The concept of angular frequency, represented by the symbol ω, is additionally introduced and defined.

$$\omega = 2\pi v$$

The unit of measurement for angular velocity is commonly denoted as radian per second (rad/s) or simply radian per second (rad/s). This is because 2π represents the angle around a circle in radians, and T represents the time period in seconds (Pambayun et al., 2019).

7.4. DIFFERENTIAL EQUATION OF SHM

In this discourse, we shall proceed to articulate the equation governing Simple Harmonic Motion (SHM) in its differential form, employing Newton's second law of motion. According to Newton's second law of motion, the force exerted on an object with mass m may be mathematically represented as a function of its acceleration,

$$F = ma = m\ddot{x}$$

Hence, in a spring-mass system, the force may be expressed as,

$$F = m\ddot{x} = -kx$$

Or we may say that

$$m\ddot{x} + kx = 0$$

It is recommended to adhere to either the double dot notation or the d2x/dt2 notation when expressing double differentiation. The notation d2x/dt2 has been employed in subsequent units. It would be more advantageous for students to adhere to the d2x/dt2 notation (Tisdell, 2019).

The aforementioned equation represents the differential equation governing simple harmonic motion (SHM). The symbol k represents the force constant, which is commonly referred to as the spring constant in the context of a spring-mass system. It possesses dimensions of $(MLT^{-2}/L^{-})=MT^{-2}$. Hence, it can be concluded that the dimension of, k/m is T^{-2} which represents the square of the reciprocal of time. The expression k/m can be substituted by ω^2.

$$\ddot{x} + \omega^2 x = 0$$

In this analysis, we aim to elucidate the physical significance of the symbol ω, which corresponds to the angular frequency previously established (Hernández-Méndez et al., 2022).

7.4.1. Solution of the Differential Equation of SHM

The second derivative of displacement ($\ddot{x}$) may be expressed as:

$$= \frac{d^2x}{dt^2} = \frac{d}{dt}\left(\frac{dx}{dt}\right)$$

When dividing and multiplying by dx in both the numerator as well as the denominator, the resulting expression is (Gauthier, 2004):

$$\ddot{x} = \frac{dx}{dt}\frac{d}{dx}\left(\frac{dx}{dt}\right)$$

It is well-established that the derivative of position with respect to time, denoted as $\dot{x}$ or dx/dt represents the velocity v Hence, the aforementioned expression can be represented in the subsequent format:

$$\ddot{x} = v\frac{d}{dx}(v)$$

$$\frac{d}{dx}\left(\frac{v^2}{2}\right) = v\frac{dv}{dx}$$

$$\ddot{x} = \frac{d}{dx}\left(\frac{v^2}{2}\right)$$

Since,

$$\frac{d}{dx}\left(\frac{v^2}{2}\right) + \omega^2 x = 0$$

$$\text{or } \frac{d}{dx}\left(\frac{v^2}{2} + \omega^2\frac{x^2}{2}\right) = 0$$

$$\therefore d(v^2 + \omega^2 x^2) = 0$$

Upon performing the process of integration on both sides of the equation, we obtain:

$v^2 + \omega^2 x^2 = cons$

We already know that on the two extremes, if the magnitude of the displacement is equals the amplitude ($x = \pm A$), the kinetic energy or the velocity is zero ($v = 0$). Using this boundary condition, we can calculate the constant (C_1).Thus, C_1isgivenby:

$(0)^2 + \omega^2(\pm A)^2 = C_1$

or$C_1 = \omega^2 A^2$

Rearranging the terms, we get

$v^2 = \omega^2(A^2 - x^2)$

or $v = \pm\omega\sqrt{}(A2 - x2)$

The aforementioned equation represents the velocity of a particle undergoing Simple Harmonic Motion (SHM). The maximum magnitude of velocity occurs at $x = 0$, which corresponds to the mean position. The expression for the maximum velocity is as follows (Li, 2023):

$|v|max = \omega A$

7.4.2. Angular Frequency of SHM

It is understood that the displacement x(t) is expected to revert back to its original value following a single time period T of the motion. Or:

x(t) = x(t+T)

It is also recognized in the field of trigonometry that the sine or cosine function exhibits periodicity, wherein it repeats itself after an increment of $2\pi\ rad$ (Triana & Fajardo, 2013). Thus,

$\omega(t + T) = \omega t + 2\pi$

Or, we obtain

$$\omega = \frac{2\pi}{T} = 2\pi\nu$$

Hence, the symbol ω represents the angular frequency as previously defined. The SI unit associated with this quantity is expressed as "radians per second"($rads^{-1}$) (Jafarov et al., 2020).

It can be inferred that

$\omega^{2=}$

$=$

7.5. DIFFERENT KINDS OF SPRING-MASS SYSTEM

The spring-mass system has been previously discussed in the preceding sections. In the following discourse, we shall engage in an examination of various spring-mass systems (Ebrahimi et al., 2019).

7.5.1. Horizontal Oscillations

In this scenario, a spring devoid of weight is affixed firmly to a wall at one extremity, while the other extremity is connected to a body that is unrestricted in its movement on a horizontal surface devoid of friction. When an external force is applied to the body, causing it to move a short distance, the spring undergoes a deformation from its initial equilibrium state. Consequently, the body experiences a restoring force that is directly proportional to the magnitude of its displacement, leading it to initiate simple harmonic motion (SHM) (Rebillard et al., 1992).

It has also been identified that the angular frequency of simple harmonic motion (SHM) is expressed as:

$\omega=$

And the time period is expressed as:

T= 2

The displacement x as a function of time is of the form

$x(t)=A\sin(\omega t+ð)$

7.5.2. Vertical Oscillations

In this discourse, we shall now direct our attention toward the oscillatory behavior exhibited by a system comprising a vertical spring and a mass. In this scenario, we examine a spring with negligible weight and a spring constant denoted as k.

Did You Know?

The frequency of oscillation for a mass-spring system in vertical motion depends on both the spring constant and the effective mass of the system.

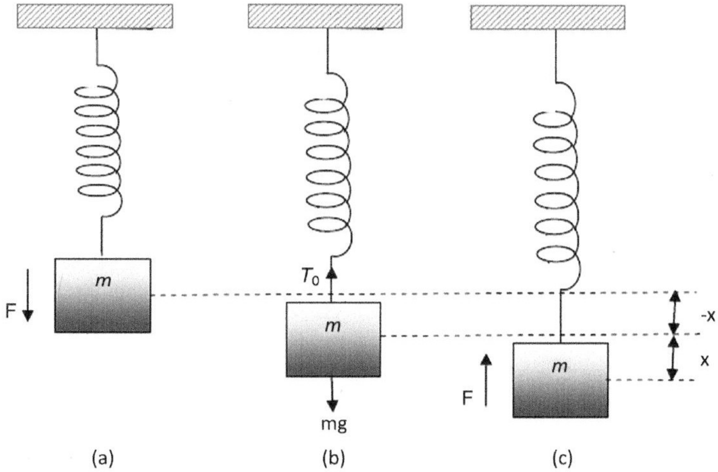

Figure 7.5. The phenomenon of oscillations in a spring that is subject to a load can be observed in three distinct configurations: (a) when the spring is compressed, (b) when it is in a state of equilibrium, and (c) when it is stretched (Source: Agrawal, Creative Commons License).

The spring is suspended in a vertical orientation from a stationary point, and it bears the weight of an object with mass m at its lower extremity, as depicted in Figure 7.5 (Qiao et al., 2002).

If the object is in a state of rest and located at its initial equilibrium position (as shown in Figure 7.5b), the tension in the spring at equilibrium, denoted as, is equivalent in magnitude but reverse in direction to the weight of the object, represented by $F = mg$ and g denotes the acceleration due to gravity, which is approximately equal to 9.8 ms^{-2}. Therefore, it can be concluded that the object is not experiencing any net force. This can be further supported by the principle of force balance, which allows us to express this relationship (Qu et al., 2019).

$$T_0 = mg$$

The tension in the spring undergoes variation as a consequence of the oscillatory motion of the object, which in turn leads to alterations in the length of the spring. If the object is displaced by a small distance x from the point of equilibrium, the force that restores it is generated by the change in tension, denoted as ΔT, from the equilibrium state. The application of Hooke's law to this alteration in tension yields (Valle et al., 2019):

$$\Delta T = -kx$$

The negative sign in the equation indicates that the change in tension occurs in the reverse direction of the displacement. Consequently, a spring under load is commonly referred to as exhibiting simple harmonic motion.

The angular frequency can be expressed as:

$\omega =$

And the time period by

T= 2

$T = 2\pi\sqrt{m/k}$

which is the same as that for horizontal oscillations?

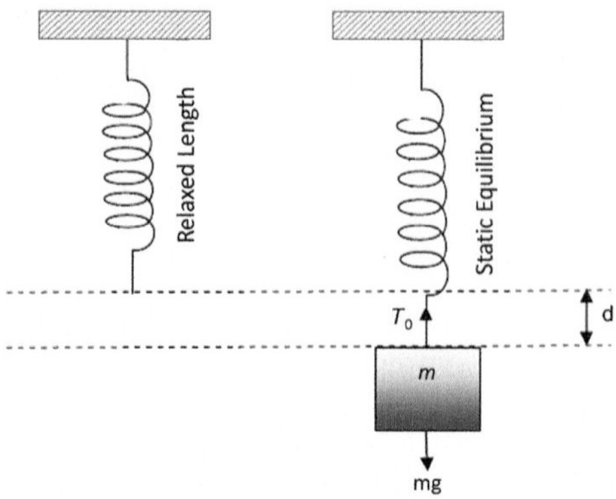

Figure 7.6. Relaxed spring hanging from the ceiling and a loaded spring (Source: Kamal Devlal, Creative Commons License)

The determination of the spring constant k in vertical oscillations is:

$T_0 = mg$

In this context, the tension denoted as T_0 is responsible for elongating the spring and can be expressed as (Shen et al., 2008):

$T_0 = -kd$

The variable "d" represents the disparity between the relaxed length as well as the equilibrium length of the spring, as depicted in Figure 7.6. Consequently, the result is as follows:

k=

where F represents the weight of the object, and it is equivalent to m multiplied by g (Xiao et al., 2012).

7.5.3. Composite Spring-Mass System

In this section, we will examine a selection of significant composite spring-mass systems. The initial item under consideration is the one which was presented to you as a question intended to test your abilities (SAQ 8). Let us endeavor to ascertain the comprehensive solution to the issue at hand. In the given illustration, depicted in Figure 7.7, there exist two springs characterized by their respective spring constants k_1 as well as k_2. These springs are connected to a solid wall at one end and to a block with mass m at the other end (Roozen et al., 2021).

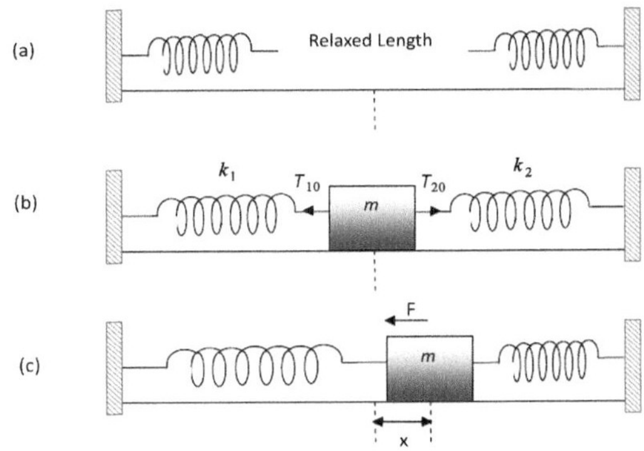

Figure 7.7. (a) The springs are affixed to the rigid wall on one side. (b) The spring-mass system is in a state of static equilibrium. (c) The spring-mass system exhibits oscillatory behavior (Source: Garg, Creative Commons License)

The determination of the spring constant for a given spring can be achieved by achieving an equilibrium of forces. Let d_1 and d_2 represent the magnitudes of the displacements of the two springs in the combined spring-mass system when it is in a state of static equilibrium, as depicted in Figure 7.7. Under this circumstance, the forces exerted by the two springs will reach equilibrium. Therefore, it is possible to formulate the following statement (Ghadiri et al., 2015):

T10=T20

or$k_1 d_1 = k_2 d_2$

If the block is displaced in the positive direction by a small distance x and subsequently released, the system undergoes Simple Harmonic Motion (SHM). The block experiences a restoring force, denoted as F, which is determined by the combined effect of the change in tension in the two springs. This is because the spring on the left is stretched, resulting in a force that attempts to pull the block in the leftward direction. Similarly, the spring on the right is compressed, generating a force that also endeavors to push the block toward the left. Therefore, it can be observed that both springs exert a concurrent force on the block, aiming to move it in an identical direction (Malekzadeh et al., 2006).

Therefore, it is possible to express, utilizing Newton's second law,

$F = \Delta T1 + \Delta T2$

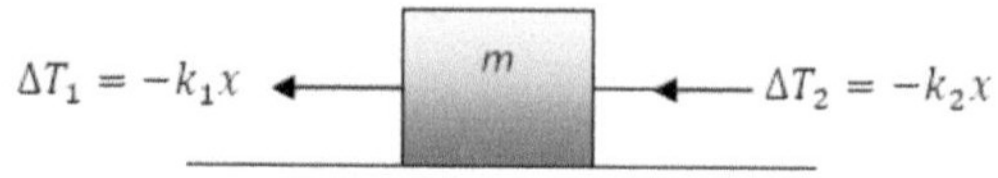

$\therefore F = -k_1 x + -k_2 x$

$= -(k_1 + k_2)x$

Therefore, it can be concluded that the behavior of this combined spring-mass system is analogous to that of the simple horizontal spring-mass system. The net spring constant of the system as a whole can be determined by adding together the spring constants of the two individual springs (Singh & Mahajan, 2016).

$k = k_1 + k_2$

7.6. ENERGY IN SIMPLE HARMONIC MOTION

7.6.1. Kinetic Energy and Potential Energy in SHM

Within the domain of oscillatory motion, a crucial aspect to investigate pertains to the interaction among kinetic energy (KE) as well as potential energy (PE). In the context of simple harmonic motion (SHM), which is a special type of periodic motion, there is a continuous transformation of two forms of energy as the oscillating system undergoes back-and-forth motion within its equilibrium position.

Kinetic Energy (KE): Kinetic energy refers to the form of energy that is inherently linked to the movement of an object. Within the framework of simple harmonic motion (SHM), let us examine a mass-spring system as an illustrative instance. When the mass undergoes displacement from its state of equilibrium, it acquires velocity as it experiences acceleration toward the point of equilibrium. The velocity reaches its maximum value if the mass passes throughout the equilibrium position. At this juncture, the kinetic energy has reached its peak value. Nevertheless, when the mass undergoes displacement from the equilibrium position, its velocity diminishes, resulting in a concomitant reduction in its kinetic energy. In contrast, as the mass reverts back to its equilibrium position, its velocity once again attains a peak value, thereby resulting in the maximum kinetic energy. Therefore, in the context of Simple Harmonic Motion (SHM), the kinetic energy undergoes oscillations, alternating between its highest and lowest magnitudes (Prayaga, 2007).

Potential Energy (PE): In contrast, potential energy refers to the energy that is linked to the position or arrangement of an object inside a force field. In the context of simple harmonic motion (SHM), when the mass moves from its equilibrium position, it undergoes a restoring force as a consequence of Hooke's law. The maximum potential energy of the mass is attained when it reaches its maximum movement from the equilibrium position. As the mass undergoes displacement toward the equilibrium position, the potential energy experiences a corresponding decrease, ultimately reaching its minimum value when the mass is precisely located at the equilibrium point. The potential energy subsequently rises as the mass undergoes motion in the opposite direction, away from the position of equilibrium, and once again attains its highest value at the maximum displacement. The observed pattern of oscillation in potential energy aligns with the expected behavior of the system in Simple Harmonic Motion (SHM) (Getachew, 2018).

7.6.2. Total Mechanical Energy of an Oscillating System

The aggregate mechanical energy (E) of a system undergoing simple harmonic motion (SHM) is the combination of its kinetic energy (KE) as well as potential energy (PE):

$$E = KE + PE$$

As previously mentioned, the potential and kinetic energies exhibit periodic variations in the context of Simple Harmonic Motion (SHM). At the point of greatest movement from the point of equilibrium, the mass exhibits zero kinetic energy, as its velocity is zero. However, the potential energy of the system reaches its maximum value. In contrast, at the point of equilibrium, the potential energy of the mass is reduced to zero due to its displacement being zero, while the kinetic energy reaches its highest value. Hence, the overall mechanical energy of the system stays conserved throughout the entirety of the motion. Throughout simple harmonic motion (SHM), the total mechanical energy stays constant, thereby exemplifying the principle of energy conservation within the system. One notable attribute of Simple Harmonic Motion (SHM) that distinguishes it from other forms of oscillations is its inherent resistance to energy dissipation caused by damping and external forces (Maamer et al., 2019).

7.6.3. Conservation of Energy in SHM

The concept of conserving mechanical energy in simple harmonic motion (SHM) can be comprehended through the application of the work-energy theorem. During the oscillation of the mass-spring system, the spring exerts a force that performs work on the mass, resulting in the conversion of potential energy into kinetic energy as well as vice versa. Nevertheless, the network performed during a full cycle of oscillation is nullified, since the system reverts back to its original state (Elamathi et al., 2016).

To provide an illustrative example, let us consider the following scenario:

1. At the point of equilibrium, the displacement is null, resulting in a potential energy of zero. However, the kinetic energy reaches its maximum value.
2. At the point of maximum movement, the potential energy reaches its highest value, while the velocity, and consequently the kinetic energy, becomes zero.

From these two points, there is a gradual conversion of potential energy into kinetic energy. At any given intermediate point, the total energy, which is the sum of potential as well as kinetic energies, remains constant. The perpetual conversion of energy among kinetic and potential states is a persistent process, while the overall energy within the system remains constant (Yusuf et al., 2017).

7.6.4. Phase Diagrams and Energy Exchange in Oscillatory Systems

Phase diagrams are graphical depictions employed to visually represent the correlation between two variables in relation to their respective phases, specifically their relative positions within their respective cycles. Phase diagrams are a useful tool in the study of Simple Harmonic Motion (SHM) as they provide a visual representation of the variations in the potential and kinetic energies over a complete oscillation cycle.

The phase diagram for Simple Harmonic Motion (SHM) depicts the relationship between displacement or position on the x-axis and energy on the y-axis. The shape of the potential energy curve bears a resemblance to that of a parabola, whereas the kinetic energy curve exhibits a sinusoidal pattern. At the point of equilibrium, the potential energy reaches its maximum value, while the kinetic energy becomes zero. As the system undergoes displacement towards a particular direction, the potential energy of the system diminishes, whereas the kinetic energy of the system concurrently augments. At the point of maximum displacement, the system exhibits zero potential energy and maximum kinetic energy. A symmetrical pattern is observed on the opposing side of the equilibrium position. The phase diagram provides a visual representation of the interplay among potential and kinetic energy as a system undergoes simple harmonic motion (SHM) (Stier & Nolting, 2011).

Activity 1.1

Imagine you are designing a suspension system for a car to ensure a smooth ride for passengers. The system involves a spring-mass arrangement. How will you determine the appropriate spring constant and mass to achieve the desired frequency of oscillation for the suspension? Consider factors like comfort, stability, and road conditions in your decision-making process.

The examination of energy within the context of simple harmonic motion is of utmost importance in comprehending the dynamics of oscillatory systems. The dynamic interaction between kinetic and potential energies gives rise to a perpetual exchange of energy, thereby ensuring the preservation of total mechanical energy throughout the motion.

The preservation of energy is an inherent and essential attribute of Simple Harmonic Motion (SHM), rendering it a captivating and pivotal subject within the realms of physics and engineering. By acquiring an understanding of the fundamental principles governing energy in oscillatory systems, we can obtain valuable insights into a wide range of real-world phenomena. Furthermore, this knowledge empowers us to develop efficient systems that have practical applications (Löfwander et al., 2007).

SUMMARY

This chapter has focused on the concept of periodic motion, oscillatory motion, and the conditions and fundamental characteristics of simple harmonic motion (SHM). The topic of our study pertained to the phenomenon of the restoring force, which arises as a consequence of movement from the mean or equilibrium position. Specifically, we examined how this restoring force is directly proportional to the amount of displacement in the context of simple harmonic motion. In our investigation, we examined two elementary systems, namely the simple pendulum and the spring-mass system, both of which serve as illustrations of simple harmonic motion (SHM). It is currently understood that the oscillation of a physical system is attributed to the properties of inertia as well as elasticity. By applying Newton's second law of motion, we formulated the equation of motion for Simple Harmonic Motion (SHM) and obtained the solution to the corresponding differential equation that describes SHM. In addition, we conducted calculations to determine the time period as well as angular frequency of a spring-mass system, as well as for certain composite spring-mass systems.

REVIEW QUESTIONS

1. What is simple harmonic motion (SHM), and what are its basic characteristics?
2. Describe the differential equation of SHM and how it relates to the angular frequency of oscillation.
3. Compare and contrast the motion of a simple pendulum and a spring-mass system in SHM.
4. How does the total mechanical energy of an oscillating system remain conserved in SHM?
5. Explain the concept of phase diagrams and how they represent energy exchange in oscillatory systems.
6. In a vertical oscillation of a spring-mass system, what factors determine the frequency of oscillation, and how does it differ from horizontal oscillations?

REFERENCES

1. Adolphus, T., Alamina, J., & Aderonmu, T. S. (2013). The effects of collaborative learning on problem solving abilities among senior secondary school physics students in simple harmonic motion. *Journal of Education and Practice*, *4*(25), 95–100.
2. Andrews, D. L., & Romero, L. C. D. (2009). A back-to-front derivation: the equal spacing of quantum levels is a proof of simple harmonic oscillator physics. *European Journal of Physics*, *30*(6), 1371.
3. Brodie, M. A., Walmsley, A., & Page, W. (2008). Dynamic accuracy of inertial measurement units during simple pendulum motion. *Computer Methods in Biomechanics and Biomedical Engineering*, *11*(3), 235–242.
4. Earley, J. J. (2023). Thermogenic regulation of the simple harmonic motion of tropomyosin by weak-binding myosin cross-bridges explains spontaneous oscillatory contractions. *Biophysical Journal*, *122*(3), 118a-119a.
5. Ebrahimi, F., Hajilak, Z. E., Habibi, M., & Safarpour, H. (2019). Buckling and vibration characteristics of a carbon nanotube-reinforced spinning cantilever cylindrical 3D shell conveying viscous fluid flow and carrying spring-mass systems under various temperature distributions. *Proceedings of the Institution of Mechanical Engineers, Part C: Journal of Mechanical Engineering Science*, *233*(13), 4590–4605.
6. Elamathi, N., Nusrath, R., & Jayanthi, S. (2016). An Energy Aware Fault Tolerance Scheme with Delay Reduction in WSN for Structural Health Monitoring.
7. Evans, J. W., & Michard, R. (1962). Observational Study of Macroscopic Inhomogeneities in the Solar Atmosphere. III. Vertical Oscillatory Motions in the Solar Photosphere. *Astrophysical Journal, vol. 136,* p. 493, 136, 493.
8. Festiana, I., Herlina, K., Kurniasari, L. S., & Haryanti, S. S. (2019). Damping Harmonic Oscillator (DHO) for learning media in the topic damping harmonic motion. In *Journal of Physics: Conference Series* (Vol. 1157, No. 3, p. 032062). IOP Publishing.
9. Gauthier, N. (2004). Novel approach for solving the equation of motion of a simple harmonic oscillator. *International Journal of Mathematical Education in Science and Technology*, *35*(3), 446–452.

10. Getachew, G. (2018). A review of simple harmonic motion for mass spring system and its analogy to the oscillations in lc circuit. *J Phys Chem Sci*, *6*(3).
11. Ghadiri, M., Malekzadeh, K., & Ghasemi, F. A. (2015). Free Vibration of an Axially Preloaded Laminated Composite Beam Carrying a Spring-Mass-Damper System with a Non-Ideal Support. *Jordan Journal of Mechanical & Industrial Engineering*, *9*(3).
12. Gürgöze, M. (1996). On the eigenfrequencies of a cantilever beam with attached tip mass and a spring-mass system. *Journal of Sound and Vibration*, *190*(2), 149–162.
13. Hernández-Méndez, R., Cuevas-Vallejo, A., & Orozco-Santiago, J. (2022). Didactic proposal to re-signify the differential equation by modelling simple harmonic motion with the support of digital technology. In *Fourth conference of the International Network for Didactic Research in University Mathematics*.
14. Jafarov, E. I., Nagiyev, S. M., Oste, R., & Van der Jeugt, J. (2020). Exact solution of the position-dependent effective mass and angular frequency Schrödinger equation: harmonic oscillator model with quantized confinement parameter. *Journal of Physics A: Mathematical and Theoretical*, *53*(48), 485301.
15. Lazos, P., Kyriazopoulos, N., & Nezis, A. (2022). Demonstrating the connection between uniform circular motion and simple harmonic motion with a 3D-printed model. *Physics Education*, *57*(2), 023006.
16. Li, J. (2023). Using Differential Equation to Explain the Simple Harmonic Motion Equation. *Highlights in Science, Engineering and Technology*, *49*, 438–447.
17. Lima, F. M. S., & Arun, P. (2006). An accurate formula for the period of a simple pendulum oscillating beyond the small angle regime. *American Journal of Physics*, *74*(10), 892–895.
18. Löfwander, T., Champel, T., & Eschrig, M. (2007). Phase diagrams of ferromagnet-superconductor multilayers with misaligned exchange fields. *Physical Review B*, *75*(1), 014512.
19. Maamer, B., Boughamoura, A., El-Bab, A. M. F., Francis, L. A., & Tounsi, F. (2019). A review on design improvements and techniques for mechanical energy harvesting using piezoelectric and electromagnetic schemes. *Energy Conversion and Management*, *199*, 111973.

20. Malekzadeh, K., Khalili, M. R., & Mittal, R. K. (2006). Analytical prediction of low-velocity impact response of composite sandwich panels using new TDOF spring–mass–damper model. *Journal of Composite Materials*, *40*(18), 1671–1689.
21. Musik, P. (2017). Development of Computer-Based Experiment Set on Simple Harmonic Motion of Mass on Springs. *Turkish Online Journal of Educational Technology-TOJET*, *16*(4), 1–11.
22. Pambayun, B., Wirjawan, J. V., Wijaya, A., Untung, G. B., & Pratidhina, E. (2019). Designing mobile learning app to help high school students to learn simple harmonic motion. *International Journal on Social and Education Sciences*, *1*(1), 24–29.
23. Prayaga, C. (2007). Interactive Simulation of Simple Harmonic Motion with VPython and Excel Controlled by VBA. In *EdMedia+ Innovate Learning* (pp. 3744–3748). Association for the Advancement of Computing in Education (AACE).
24. Qiao, H., Li, Q. S., & Li, G. Q. (2002). Vibratory characteristics of flexural non-uniform Euler–Bernoulli beams carrying an arbitrary number of spring–mass systems. *International Journal of Mechanical Sciences*, *44*(4), 725–743.
25. Qu, Y., Peng, Z., Zhang, W., & Meng, G. (2019). Nonlinear vibro-acoustic behaviors of coupled sandwich cylindrical shell and spring-mass-damper systems. *Mechanical Systems and Signal Processing*, *124*, 254–274.
26. Rebillard, E., Laulagnet, B., & Guyader, J. L. (1992). Influence of an embarked spring-mass system and defects on the acoustical radiation of a cylindrical shell. *Applied Acoustics*, *36*(2), 87–106.
27. Reyes, M. G. D. L. S. (2019). *Study of Second-Year Physics Students' Scientific Reasoning Ability and Knowledge of Simple Harmonic Motion through Guided Inquiry* (Doctoral dissertation, Prince of Songkla University).
28. Roozen, N. B., Urban, D., Piana, E. A., & Glorieux, C. (2021). On the use of dynamic vibration absorbers to counteract the loss of sound insulation due to mass-spring-mass resonance effects in external thermal insulation composite systems. *Applied Acoustics*, *178*, 107999.
29. Saputri, S. R., Wati, M., & Misbah, M. (2021). Simple harmonic motion electronic teaching materials based on authentic learning to train students' problem-solving skills: aspects of validity. In *Journal*

of Physics: Conference Series (Vol. 2126, No. 1, p. 012016). IOP Publishing.

30. Shen, J. K., Matuszewski, B. J., Shark, L. K., Skalski, A., Zielinski, T., & Moore, C. J. (2008). Deformable image registration-A critical evaluation: Demons, b-spline FFD and spring mass system. In *2008 Fifth International Conference BioMedical Visualization: Information Visualization in Medical and Biomedical Informatics* (pp. 77–82). IEEE.
31. Singh, H., & Mahajan, P. (2016). Analytical modeling of low velocity large mass impact on composite plate including damage evolution. *Composite Structures*, *149*, 79–92.
32. Stier, M., & Nolting, W. (2011). Carrier-mediated interlayer exchange, ground-state phase diagrams, and transition temperatures of magnetic thin films. *Physical Review B*, *84*(9), 094417.
33. Tisdell, C. C. (2019). An accessible, justifiable and transferable pedagogical approach for the differential equations of simple harmonic motion. *International Journal of Mathematical Education in Science and Technology*, *50*(6), 950–959.
34. Triana, C. A., & Fajardo, F. (2011). The influence of spring length on the physical parameters of simple harmonic motion. *European Journal of Physics*, *33*(1), 219.
35. Triana, C. A., & Fajardo, F. (2013). Experimental study of simple harmonic motion of a spring-mass system as a function of spring diameter. *Revista Brasileira de Ensino de Física*, *35*.
36. Valle, J., Fernández, D., & Madrenas, J. (2019). Closed-form equation for natural frequencies of beams under full range of axial loads modeled with a spring-mass system. *International Journal of Mechanical Sciences*, *153*, 380–390.
37. Vogt, P., & Kuhn, J. (2012). Analyzing simple pendulum phenomena with a smartphone acceleration sensor. *The Physics Teacher*, *50*(7), 439–440.
38. Wang, L., Ding, Z., Meng, S., Zhao, H., & Song, H. (2017). Kinematics and dynamics of a particle on a non-simple harmonic vibrating screen. *Particuology*, *32*, 167–177.
39. Womersley, J. R. (1955). XXIV. Oscillatory motion of a viscous liquid in a thin-walled elastic tube—I: The linear approximation for long waves. *The London, Edinburgh, and Dublin Philosophical Magazine and Journal of Science*, *46*(373), 199–221.

40. Xiao, Y., Wen, J., & Wen, X. (2012). Flexural wave band gaps in locally resonant thin plates with periodically attached spring–mass resonators. *Journal of Physics D: Applied Physics*, *45*(19), 195401.
41. Yusuf, A., Ishafit, I., & Indratno, T. K. (2017). Amusement Park Physics in Panggon Lunjak (Trampoline): Analysis of Kinematics and Energy Use Video Tracking. In *International Journal of Science and Applied Science: Conference Series* (Vol. 2, No. 1, pp. 32–37).

CHAPTER 8

Universal Gravitation and Planetary Motion

Contents

8.1. INTRODUCTION

The field of universal gravitation possesses a significant historical context that can be traced back to ancient civilizations. The ancient civilizations of Babylon and Egypt demonstrated an astute awareness of the consistent patterns exhibited by celestial bodies, thereby acknowledging the profound importance of these celestial phenomena within the context of their cultural and religious ideologies. Nevertheless, it was the ancient Greeks who initiated the development of basic models in order to elucidate the phenomenon of planetary motion (Borowski, 2013).

Claudius Ptolemy, a prominent figure in early astronomical discourse, presented a significant contribution in the form of the geocentric model during the 2nd century AD. Based on this particular model, Earth was regarded as the central point of the universe, with all celestial entities, such as planets and stars, believed to revolve in circular trajectories around it. Although the geocentric model successfully explained numerous observed phenomena, it necessitated intricate arrangements of epicycles in order to account for deviations from ideal circular motion, resulting in a complicated framework.

The modern comprehension of gravitation was established during the ages of 16 and 17, following the emergence of significant breakthroughs and profound insights in this field. Nicolaus Copernicus advocated for the heliocentric model, that posited the Sun as the central celestial body within the solar system. The emergence of this paradigm shift posed a challenge to the prevailing geocentric perspective, offering a more streamlined account of planetary motion (Nauenberg, 2005).

8.2. NEWTON'S LAW OF UNIVERSAL GRAVITATION

The study of gravitation underwent a significant transformation in the late 17th century due to the pioneering contributions of Isaac Newton, a renowned English mathematician, physicist, and astronomer. His seminal work, "Philosophiae Naturalis Principia Mathematica" (Mathematical Principles of Natural Philosophy), commonly referred to as the Principia played a pivotal role in this revolution (Figure 8.1). This seminal publication, which was released in the year 1687, established the fundamental principles of classical mechanics and presented a comprehensive elucidation of the concept of universal gravitation (Hsiang & Straume, 2014).

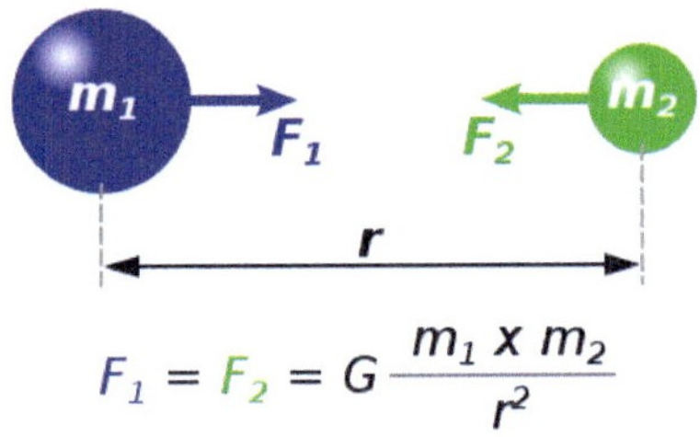

Figure 8.1. Newtons law of universal gravitation (Source: Dennis, Creative Commons License)

The Law of Universal Gravitation, formulated by Sir Isaac Newton, postulates that all particles within the cosmos exert a mutually attractive force upon one another. This force is directly related to the product of their respective masses as well as inversely correlated to the square of the distance separating them. The equation may be represented as:

F = G * (m1 * m2) / r^2

The gravitational force (F) represents the interaction among two masses (m1 as well as m2). The gravitational constant, denoted as G, is estimated to be about 6.674 × 10^-11 N m^2/kg^2. The variable "r" represents the distance separating the corresponding centers of the two masses.

This legislation consolidated the principles of celestial as well as terrestrial mechanics, demonstrating the equivalence of the force responsible for objects falling on Earth and the force that maintains the planets in their orbits surrounding the Sun. Newton's law not only offered a refined elucidation for Kepler's laws of planetary motion but also expanded the comprehension of gravitational forces to encompass all celestial entities within the cosmos (Such, 1990).

8.3. GRAVITATIONAL CONSTANT: G AND ITS SIGNIFICANCE

The gravitational constant, symbolized as G, is an intrinsic constant of the natural world that is featured in Newton's Law of Universal Gravitation. The concept is referred to denotes the magnitude of the gravitational force exerted between two objects of specified mass at a specified separation. The numerical value assigned to the gravitational constant, denoted as G, is estimated to be about 6.674 × 10^-11 N m^2/kg^2 (Papaxanthis et al., 1998).

Tip

G and its Significance (Heading number 2): "Always consider the value of the Gravitational Constant 'G' as a fixed quantity when calculating gravitational forces, as it is a universal constant and does not change."

The gravitational constant holds great significance due to its pivotal role in connecting the realms of macroscopic and microscopic phenomena.

The utilization of this concept enables the quantitative characterization of the gravitational force across various scales, encompassing the gravitational interaction between celestial bodies as well as the interplay between subatomic particles.

The utilization of G allows for the comparison of the gravitational force with other basic forces, namely electromagnetism, as well as the strong and weak nuclear forces (Schmöle et al., 2016).

Furthermore, the gravitational constant assumes a pivotal role in comprehending the configuration and progression of the cosmos. The rate of expansion of the universe is determined by the cosmological constant, a dynamic quantity introduced in Einstein's theory of general relativity to replace the gravitational constant (Kononogov & Mel'nikov, 2005).

8.4. GRAVITATIONAL FORCE AND CHARACTERISTICS

The gravitational force is a fundamental force in nature that is accountable for the attractive attraction of masses. The phenomenon under consideration is the fundamental force that regulates the movement of celestial entities, including planets, moons, stars, as well as galaxies, alongside terrestrial objects situated on the Earth's surface (Figure 8.2).

The phenomenon of objects falling to the ground when released and the maintenance of planetary orbits around the Sun can be attributed to the gravitational force (Tagawa et al., 2001).

As per Newton's Law of Universal Gravitation, the gravitational force exerted between two objects is directly proportional to the multiplication of their masses as well as inversely proportional to the square of the distance separating their centers. In a mathematical context, it can be represented as:

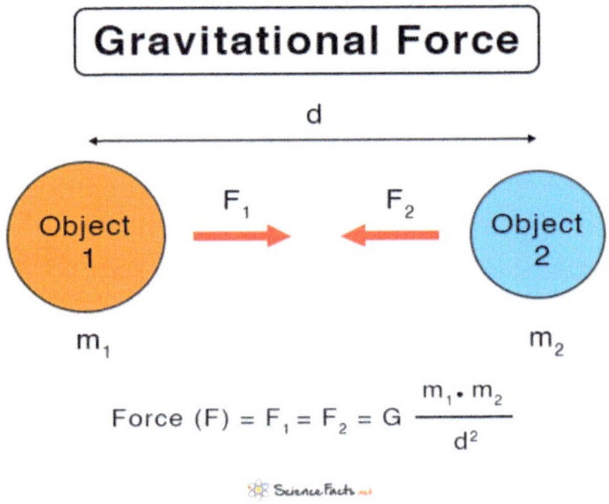

Figure 8.2. Schematic of the gravitational force (Source: Science Facts, Creative Commons License)

$F = G * (m1 * m2) / r^2$

The gravitational force, denoted as F, represents the interaction between two masses, m1 and m2. The gravitational constant, denoted as G, is approximately equal to 6.674 × 10^-11 N m^2/kg^2. The variable "r" represents the distance separating the centers of the masses.The gravitational force is consistently attractive, exerting a pulling effect on objects, causing them to move toward one another along the line that connects their respective centers. The magnitude of the gravitational force is contingent upon the masses of the entities in question as well as the separation distance between them. The magnitude of the force is directly proportional to the mass of the objects involved, such that larger masses correspond to a greater force. Conversely, the force weakens as the distance between the masses increases (Shigemitsu et al., 2003).

8.4.1. Inverse Square Law: Implications and Explanations

The inverse square law, as stated in Newton's Law of Universal Gravitation, holds considerable implications for the dynamics of gravitational force. As per the stipulations of this law, the magnitude of the force experienced between two masses diminishes proportionally to the square of the distance separating them. This phenomenon has numerous significant implications.

1. Decreasing Strength with Distance: As the distance between two masses increases, the gravitational force exerted between them diminishes significantly. When the distance among the masses is

increased by a factor of two, the force between them decreases by a factor of four. Similarly, when the distance is increased by a factor of three, the force decreases by a factor of nine. This relationship holds for further increases in distance. This particular attribute guarantees that gravitational interactions diminish in significance over vast distances, thereby enabling other fundamental forces to exert greater influence on larger scales (Vatansever et al., 2022).

2. Orbits of Planets and Satellites: The motion of planets around the Sun as well as satellites around planets is governed by the inverse square law. According to this law, as a planet or satellite approaches the celestial body it orbits, the gravitational force exerted on it increases, resulting in an acceleration that leads to an increase in its orbital speed. Conversely, as the planet or satellite moves farther away from the celestial body, the gravitational force weakens, causing a deceleration and a decrease in orbital speed. This delicate equilibrium between gravitational attraction as well as the object's inertial motion gives rise to stable elliptical orbits (Vezzoli, 2001).
3. Gravitational Waves: The inverse square law is of significant importance in the transmission of gravitational waves. The generation of these ripples in the fabric of spacetime is attributed to the acceleration of massive celestial bodies, such as the merger of black holes and neutron stars. Gravitational waves exhibit a phenomenon wherein their intensity diminishes as they propagate through space, owing to the inverse square law (Sannikova, 2021).

8.4.2. Gravitational Force between Multiple Bodies

When confronted with multiple masses, the gravitational force that affects a specific mass is the resultant vector of the individual forces generated by all other masses. In the context of a two-body system, such as the Earth-Moon system, it is observed that the gravitational force exerted on Earth can be expressed as the resultant vector of the gravitational forces exerted by both the Moon and the Sun.

In a trinary celestial system, like the Sun-Earth-Moon system, the gravitational interactions exhibit heightened complexity. The gravitational force exerted by the Sun is the primary factor governing the orbital paths of both the Earth as well as the Moon. However, it is worth noting that the gravitational force exerted by the Moon also exerts an influence on the Earth's motion. The gravitational force exerted by the Moon induces tidal phenomena on Earth, resulting in the periodic elevation and recession of oceanic tides (Johnson & Litehiser, 1972).

In systems comprising more than three celestial bodies, like the solar system, the gravitational forces exerted by planets, moons, and other celestial bodies play a significant role in shaping the complex dynamics of orbital motion and interactions. Comprehending intricate gravitational interactions necessitates the utilization of sophisticated mathematical methodologies, such as numerical simulations as well as perturbation theory (Lopez, 2007).

8.5. KEPLER'S LAWS OF PLANETARY MOTION

During the early 17th century, Johannes Kepler, a renowned German astronomer as well as mathematician, conducted significant observations and analyses pertaining to planetary motion, which were considered groundbreaking in the field.

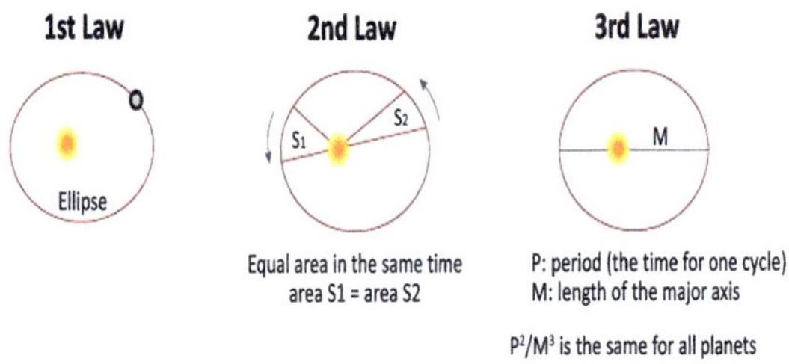

Figure 8.3. Kepler's laws of ellipses (Source: Abhi Patel, Creative Commons License)

The contributions of this individual significantly transformed our comprehension of celestial mechanics, ultimately laying the foundation for the development of Isaac Newton's law of universal gravitation. Kepler's Laws of Planetary Motion (Figure 8.3), a set of three principles, were formulated by Johannes Kepler to elucidate the dynamics of planetary motion in relation to the Sun (Abdel-Basset et al., 2023).

8.5.1. Kepler's First Law: Law of Ellipses

According to Kepler's First Law, the trajectory of a celestial body, such as a planet, as it revolves around the Sun follows an elliptical path, wherein the

Sun occupies one of the two focal points of the ellipse. An ellipse can be defined as a closed curve wherein the total distance from any given point on the curve to both foci remains constant. Within the framework of planetary motion, it is observed that the Sun occupies a position at one of the foci, while the other focus remains unoccupied (Hsiang et al., 2015).

The primary characteristics of an ellipse are its major axis and minor axis. The major axis of an ellipse refers to the longest diameter that passes through its center. It is precisely two times the length of the semi-major axis, which is commonly denoted as 'a'. The semi-major axis is defined as one-half of the major axis as well as serves to determine the mean distance between a planet and the Sun.

The implications of Kepler's First Law on the nature of planetary orbits are significant. Before the time of Kepler, the dominant belief was that the trajectories of planets followed perfectly circular paths. Nevertheless, Kepler's research established that planetary orbits exhibit an elliptical shape, characterized by fluctuating distances between the celestial body and the Sun at different stages of its orbital path. The aforementioned development marked a notable divergence from preceding models, resulting in a heightened precision in the depiction of planetary motion (Marcacci, 2022).

8.5.2. Kepler's Second Law: Law of Equal Areas

Kepler's Second Law, commonly known as the Law of Equal Areas, posits that the orbital motion of a planet around the Sun results in the sweeping out of equal areas along the trajectory within equal time intervals. Put simply, the velocity of a planet during its orbital motion is not uniform; it exhibits an increased speed when it is in proximity to the Sun, and a decreased speed when it is at a greater distance.In order to enhance comprehension of this law, it is advisable to contemplate the line segment that connects a celestial body, such as a planet, to the central star of its respective solar system, commonly referred to as the radius vector. As the celestial body traverses its elliptical trajectory, the vector connecting its center to a given point on the orbit traces out regions within the orbital plane. According to Kepler's Second Law, the angular velocity of the radius vector remains constant, resulting in a constant rate of area swept out by the vector (Krittinatham & Kaewkong, 2015).

When a planet is in proximity to the Sun, specifically during perihelion, it exhibits a higher velocity, resulting in an increased displacement over a specified time interval. In contrast, during its aphelion, when the distance between the Earth and the Sun is at its greatest, the speed of the Earth decreases, resulting in a decrease in its orbital velocity. This observation aligns with the principle of angular momentum conservation, a fundamental concept in the field of physics.

Kepler's Second Law provides a significant understanding of the mechanics involved in the motion of celestial bodies. This phenomenon elucidates the reasons behind the diverse orbital velocities exhibited by planets at different positions along their orbits, and provides insights into the seasonal fluctuations observed on Earth due to its elliptical trajectory around the Sun (Spolter, 2011).

8.5.3. Kepler's Third Law: Law of Harmonies

Kepler's Third Law, commonly referred to as the Law of Harmonies, elucidates the correlation among the orbital periods and semi-major axes of celestial bodies such as planets. The statement asserts that the square of a planet's orbital period (T) is directly correlated to the cube of its semi-major axis (a).

In mathematical terms, Kepler's Third Law can be formulated as follows:

T^2 = k * a^3

The variable T represents the orbital period of the planet, which denotes the duration required for the planet to complete a full revolution around the Sun. The semi-major axis, denoted as "a," represents the length of the major axis of the elliptical orbit of the planet. The constant of proportionality, denoted as k, remains consistent across all planets in their orbits around the Sun (Lombardi, 2018).

According to Kepler's Third Law, planets situated at greater distances from the Sun exhibit longer orbital periods in comparison to those located in closer proximity to the Sun.

The identification of this relationship constituted a significant breakthrough in the field of astronomy, enabling astronomers to ascertain the relative distances among planets within the solar system, even in the absence of knowledge regarding their absolute distances.

The Law of Harmonies played a crucial role in substantiating the heliocentric model postulated by Copernicus and presented compelling evidence contradicting the geocentric model. The laws established by Kepler, in conjunction with Newton's subsequent formulation of the law of universal gravitation, served as the foundation for contemporary studies in celestial mechanics as well as astrophysics (Molland, 1975).

8.6. DERIVING KEPLER'S LAWS FROM UNIVERSAL GRAVITATION

8.6.1. Applying Universal Gravitation to Planetary Motion

The derivation of Kepler's Laws of Planetary Motion from Newton's Law of Universal Gravitation entails the application of the gravitational force equation to the orbital motion of celestial bodies revolving around the central body, which in this case is the Sun. According to Newton's Law of Universal Gravitation, the gravitational force among a celestial body with mass M, such as the Sun, and another celestial body with mass m, such as a planet, can be mathematically represented as (Wilson, 1970):

$$F = G * (M * m) / r^2$$

The variable F represents the gravitational force exerted among the Sun and the planet. The gravitational constant, denoted as G, is a fundamental physical constant that plays a crucial role in the field of physics. The variable "r" represents the distance separating the central positions of the Sun as well as the planet.

The subsequent stage involves the examination of the centripetal force necessary to maintain the planet within its orbital path. The centripetal force for a circular orbit can be expressed as:

$$Fc = m * v^2 / r$$

The symbol Fc represents the centripetal force. The variable "m" represents the mass of the planet. The circular velocity of the planet is denoted by v (Zhao et al., 2010).

In the case of a circular orbit, the centripetal force is equivalent to the gravitational force. Therefore, by setting the equations for gravitational force as well as centripetal force equal to each other, we obtain:

$$G * (M * m) / r^2 = m * v^2 / r$$

The equation can be simplified through the elimination of the mass of the planet (m) on both sides, resulting in:

$$G * M / r^2 = v^2 / r$$

Subsequently, the equation is reorganized in order to determine the orbital velocity (v) as a function of the distance from the Sun (r).

$v^2 = G * M / r$

By applying the square root operation to both sides of the equation, we obtain:

$v = \sqrt{(G * M / r)}$

The equation presented herein demonstrates that the orbital velocity of a celestial body is contingent upon the mass of the central star, denoted as M, as well as the radial distance between said celestial body and the central star, denoted as r. The orbital velocity remains unaffected by the mass of the planet (m), indicating that planets situated at equal distances from the Sun will exhibit identical orbital velocities. This observation aligns with Kepler's Second Law, which posits that planets traverse equal areas in equal time intervals, indicating that they exhibit greater velocity in proximity to the Sun and reduced velocity at greater distances (Barker, & O'connell, 1970).

8.6.2. Mathematical Derivation of Kepler's Laws

The first law of Kepler, which asserts that the trajectory of a celestial body revolving around the Sun takes the form of an ellipse with the Sun positioned at one of the two foci, can also be deduced from Newton's Law of Universal Gravitation. In order to establish this principle, we examine the aggregate mechanical energy (E) of the celestial body within its orbital trajectory, which encompasses both its kinetic energy (KE) and gravitational potential energy (PE). The expression denoting the kinetic energy (KE) of the planet is as follows:

$KE = (1/2) * m * v^2$

The gravitational potential energy (PE) among the Sun and a planet can be expressed as:

$PE = -G * (M * m) / r$

The inclusion of a negative sign in the equation for potential energy indicates that the potential energy is of a negative value. This negative value indicates that the potential energy represents the work performed by the gravitational force when the planet is being drawn toward the Sun (Mošna, 2014). The aggregate mechanical energy (E) is the combination of both kinetic and potential energies.

$E = KE + PE$ $E = (1/2) * m * v^2 - G * (M * m) / r$

Given that the total mechanical energy (E) is conserved, it remains constant throughout all locations within the planet's orbit.

In order to examine the configuration of the trajectory, the semi-major axis (a) as well as the semi-minor axis (b) of the elliptical path is introduced. The semi-major axis represents one-half of the greatest diameter of the ellipse, whereas the semi-minor axis corresponds to one-half of the shortest diameter. The equation that relates the distance (r) among the Sun and a planet in an elliptical orbit to the semi-major axis (a) as well as the semi-minor axis (b) is as follows:

$r = a * (1 - e^2)$

The variable "e" in this context denotes the eccentricity of an ellipse, which serves as a measure of the deviation of the ellipse from a circle that is perfect It is important to note that the value of "e" falls within the range of 0 to 1, inclusively. In the case of a circle, the eccentricity (e) is equal to 0, while for an elongated ellipse, the eccentricity (e) tends to approach 1 (Baez, 1960).

By replacing the expression for the variable (r) into the equation for the total mechanical energy (E), the resulting equation is obtained:

$E = (1/2) * m * v^2 - G * (M * m) / (a * (1 - e^2))$

The aforementioned equation provides a comprehensive representation of the overall mechanical energy possessed by the planet at any given juncture within its orbital trajectory. Given that the constant (E) remains unchanged, it is possible to manipulate the equation in order to determine the square of the orbital velocity (v^2).

$v^2 = 2 * [G * (M * m) / (a * (1 - e^2)) + G * (M * m) / r]$

$v^2 = 2 * G * (M * m) * [1 / (a * (1 - e^2)) + 1 / r]$

Utilizing the established relationship between the major axis (r) and the semi-major axis (a) for an ellipse, denoted as ($r = a * (1 - e^2)$), it is possible to proceed with further simplification of the equation.

$v^2 = 2 * G * (M * m) / a$

Ultimately, by applying the square root operation to both sides of the equation, we arrive at the mathematical representation of the orbital velocity (v) in relation to the semi-major axis (a).

$v = \sqrt{(2 * G * (M * m) / a)}$

The aforementioned equation demonstrates a direct proportionality between the orbital velocity of a planet and the opposite of the square root of its semi-major axis. The aforementioned statement aligns with Kepler's Third Law, which posits that the square of a planet's orbital period (T) exhibits a direct relationship with the cube of its semi-major axis (a) (Setyadin et al., 2020).

8.7. ORBITAL MOTION AND ESCAPE VELOCITY

8.7.1. Circular and Elliptical Orbits

Within the domain of celestial mechanics, the concept of orbital motion pertains to the dynamic behavior exhibited by a celestial entity as it revolves around another celestial entity, driven by the force of gravitational attraction. There exist two fundamental categories of orbits, namely circular and elliptical.

1. Circular orbits can be considered as a specific instance of elliptical orbits, characterized by an eccentricity (e) value of zero. In a circular orbit, the celestial body or satellite undergoes a continuous revolution around the central mass, maintaining a consistent distance and exhibiting a stable, consistent motion. The centripetal force necessary to maintain the circular trajectory of the orbiting object is provided by the gravitational force exerted between the two bodies. An instance of an orbit that is nearly circular can be observed in the case of the Earth revolving around the Sun, wherein the eccentricity is extremely close to zero (Carney et al., 1988).
2. Elliptical Orbits: An elliptical orbit is characterized by a non-zero eccentricity (e) value, indicating that the orbit deviates from a perfect circle. In this scenario, the celestial object undergoes motion characterized by various distances from the central body, leading to non-uniform movement. The gravitational force continues to serve as the centripetal force, ensuring the object remains in orbit. However, due to the varying distances, there are disparities in the orbital speed at various points within the orbit. The planetary orbits within our solar system exhibit a predominantly elliptical shape, wherein the Sun occupies a focal point of the ellipse (Wu et al., 2008).

8.7.2. Speed and Velocity of Planets in Orbit

The determination of a planet or satellite's speed and velocity while in orbit is of paramount importance in comprehending its orbital motion within the central celestial object.

1. Speed in Orbit: The velocity of an object in orbit undergoes variation as it traverses its trajectory. According to Kepler's Second Law, also known as the Law of Equal Areas, the trajectory of a planet results in the sweeping out of equal areas in equal time intervals along the line segment connecting the planet as well as the Sun. This suggests that the velocity of the object is not uniform, but rather undergoes variations as it approaches or recedes from the central mass. In solar orbits, the velocity of an object is greater when it is in proximity to the central body (perihelion) and decreases as it moves further away (aphelion) (Kempton, & Rauscher, 2012).
2. Velocity in Orbit: Velocity is a vector quantity that incorporates both magnitude (speed) and direction. In a circular orbit, the velocity of a planet and satellite is tangentially aligned with the orbit at any given location. Consequently, the object's velocity undergoes continuous changes in direction as it traverses its elliptical trajectory, while its magnitude, or speed, fluctuates in accordance with Kepler's Second Law (Lissauer, 2007).

Remember

A planet's speed in its orbit varies; it is faster when it is closer to the sun and slower when it is farther from the sun due to the gravitational pull.

8.7.3. Escape Velocity: Concept and Importance

Escape velocity refers to the minimum velocity that an object must achieve in order to overcome the gravitational pull exerted by a celestial body, like a planet or moon.

By attaining this velocity, the object can successfully transition into space without any possibility of returning. The concept holds fundamental significance in the domains of space exploration and celestial mechanics (Figure 8.4).

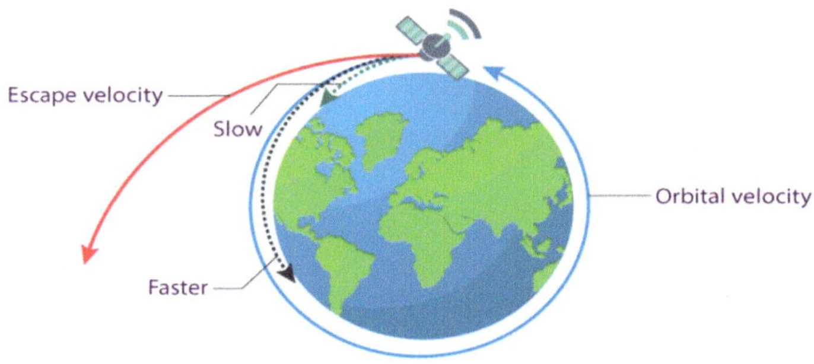

Figure 8.4. Schematic representations of the escape velocity (Source: Byju's Creative Commons License)

The escape velocity is contingent upon the mass (M) as well as the radius (R) of the celestial body to which the object is endeavoring to emancipate itself. Mathematically, the escape velocity (Vesc) may be represented as:

$$Vesc = \sqrt{(2 * G * M / R)}$$

The variable "Vesc" represents the escape velocity. The gravitational constant, denoted as G, is a fundamental physical constant in the field of physics. The variable "M" represents the mass of the celestial body. R represents the radial distance between the center of the celestial body and the initial position of the object.

The significance of escape velocity is paramount in the realm of space exploration and missions. To illustrate, in order to initiate the ascent of a spacecraft into outer space, it is imperative for it to attain a velocity surpassing the escape velocity of the Earth, thereby surpassing the gravitational force exerted by the Earth. If the spacecraft fails to achieve the necessary escape velocity, it will experience gravitational forces that will cause it to either return to Earth or enter into an orbit that is not stable. The concept of escape velocity is of great significance in comprehending the dynamics of celestial bodies within the cosmos. The comprehension of escape velocities for celestial entities such as planets and moons is crucial in ascertaining their capacity to maintain an atmosphere or retain satellites and moons (Lim et al., 2014).

The idea of escape velocity exhibits a close correlation with orbital velocity. The minimum velocity necessary for an object to maintain its position in a circular and elliptical orbit is referred to as the orbital velocity. In the event that the velocity of an object surpasses the orbital speed, it will disengage from its present orbit and transition into a higher orbit that is situated at a greater distance. Once the object attains the escape velocity, this will achieve liberation from the gravitational influence exerted by the celestial body, thereby achieving complete detachment (Ikegaya et al., 2022).

8.8. GEOSYNCHRONOUS AND HELIOCENTRIC ORBITS

8.8.1. Geosynchronous Orbits: Satellite Applications

A geosynchronous orbit refers to a distinct orbital path around the Earth in which a satellite completes one revolution in sync with the Earth's rotation. As a consequence of this phenomenon, the satellite exhibits a state of immobility in relation to a specific location on the Earth's surface, rendering it highly suitable for a multitude of satellite-based endeavors.

1. Communication Satellites: Communication satellites often utilize geosynchronous orbits due to their ability to maintain a fixed position relative to Earth, thereby enabling uninterrupted and reliable communication coverage over a designated geographic area. The provision of reliable and extensive coverage for television broadcasting, online services, as well as long-distance phone calls heavily depends on the utilization of geosynchronous communication satellites (Bruno et al., 2006).

Did You Know?

Did you know that satellites in geosynchronous orbits stay at a fixed position relative to the Earth's surface, making them invaluable for communication, weather observation, and other long-term applications?

2. Weather Satellites: The continuous observation and tracking of weather systems over a fixed region by geosynchronous satellites greatly enhance the effectiveness of weather monitoring and forecasting. Continuous monitoring is of utmost importance in order to ensure timely and precise weather forecasting and effective disaster management.
3. Earth Observation Satellites: Geosynchronous orbits find utility in the realm of Earth observation as well as remote sensing applications. These satellites possess the capability to consistently observe designated areas, thereby offering significant data for the purpose of environmental monitoring, disaster preparation, as well as land use mapping.

4. Navigation and GPS: Although not as prevalent as medium Earth orbit satellites utilized for global positioning systems (GPS), certain geosynchronous satellites serve the purpose of regional navigation as well as GPS augmentation, particularly in regions where terrestrial infrastructure is limited (Zhou et al., 2018) (Figure 8.5).

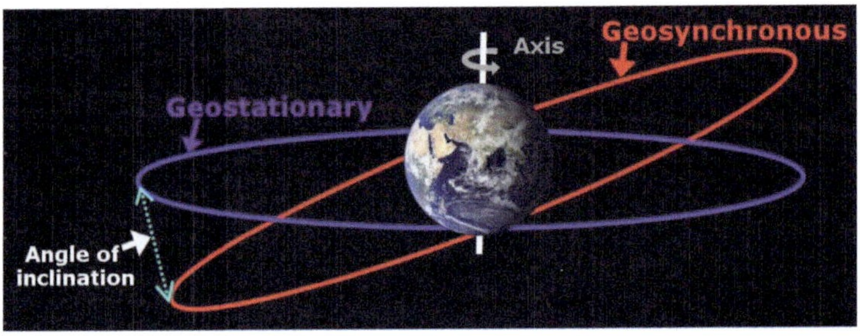

Figure 8.5. Geosynchronous vs geostationary satellites (Source: Ashish, Creative Commons License)

8.8.2. Heliocentric Orbits of Planets in the Solar System

Heliocentric orbits pertain to the celestial paths followed by planets as they revolve around the central star, known as the Sun. The Solar System serves as a prominent illustration of heliocentric orbital patterns, wherein the Sun acts as the central gravitational focus around which the primary celestial bodies, including Earth, orbit (Stern, 2005).

1. Elliptical Orbits: The laws formulated by Johannes Kepler elucidate the fundamental characteristics of planetary motion, specifically the elliptical shape of orbits within a heliocentric system. Every celestial body in our solar system, including the planets, adheres to an elliptical trajectory as it orbits around the Sun. The Sun occupies one of the two foci of the elliptical path. The orbital speeds of planets are subject to changes due to the different distances between them and the Sun, as elucidated by Kepler's Second Law (Brunini, 2023).
2. Planetary Orbital Periods: The relationship between the orbital period (T) of a planet and its mean distance (semi-major axis, a) from the Sun is described by Kepler's Third Law. The relationship between the square of the orbital period and the cube of the semi-major axis can be expressed as T^2 ∝ a^3. This

correlation enables astronomers to ascertain the comparative distances between planets within the Solar System by utilizing their respective orbital periods (Dobrovolskis et al., 2007).

3. Orbital Resonances: Planets can exhibit orbital resonances as a consequence of their gravitational interactions in certain scenarios. As an illustration, the orbital period of Jupiter is estimated to be approximately 11.86 years, whereas Saturn's orbital period is approximately 29.46 years. This phenomenon results in the establishment of a 2:5 orbital resonance, wherein Jupiter completes about two revolutions surrounding the Sun for every five revolutions of Saturn. The presence of such resonances has the potential to exert an influence on the equilibrium and long-term dynamics of planetary orbits (Trilling et al., 1998).

8.8.3. Stability and Long-Term Behavior of Orbits

The stability and long-term dynamics of orbits are contingent upon various factors, including the spatial arrangement of mass within the celestial system as well as the gravitational interactions among celestial bodies.

1. Stability of Geosynchronous Orbits: Geosynchronous orbits exhibit a notable degree of stability primarily as a consequence of their synchronous alignment with the rotational motion of the Earth. Nevertheless, minor disturbances, such as gravitational forces exerted by celestial bodies like the Moon and the Sun, as well as atmospheric resistance, can gradually impact the orbital position of a satellite. In order to sustain a geosynchronous orbit, satellite operators employ small thrusters to execute periodic adjustments (Colombo, 2015).
2. Stability of Heliocentric Orbits: The stability of heliocentric orbits is primarily attributed to the prevailing influence of the Sun's gravitational force. Nevertheless, the occurrence of planetary perturbations, specifically the gravitational interactions among planets, has the potential to induce enduring alterations in orbital parameters. The occurrences referred to as gravitational perturbations play a crucial role in various phenomena, including the rotation of planetary orbits (Zeebe, 2015).
3. Resonances and Orbital Instabilities: The stability of orbits in both geosynchronous as well as heliocentric configurations can be influenced by resonances occurring between celestial bodies. The occurrence of orbital resonances can result in cyclic fluctuations

in orbital parameters, thereby influencing the extended dynamics of celestial entities (McComas et al., 2011).

Geosynchronous orbits possess extensive utility in various domains, including communication, monitoring the weather, and Earth observation satellites. They facilitate the provision of consistent and uninterrupted coverage of particular areas on Earth. The heliocentric model elucidates the trajectory of celestial bodies within the Solar System, wherein Kepler's laws govern the elliptical nature of their paths and determine their orbital periods. The stability and long-term dynamics of orbits are contingent upon a multitude of gravitational interactions as well as resonances that occur between celestial bodies. The examination of these orbital paths and their associated dynamics is imperative in order to gain comprehension of satellite missions, space-related endeavors, and the behavior exhibited by celestial entities within the vast expanse of the universe (Lecar et al., 1992).

Activity 1.1

Imagine you are tasked with launching a satellite into geosynchronous orbit. Given the mass of the satellite, the gravitational constant, and the radius of the Earth, calculate the speed at which the satellite must move to maintain a stable geosynchronous orbit. Also, discuss the importance of this speed being maintained for the satellite's successful operation.

SUMMARY

This chapter provides a comprehensive understanding of the principles governing the motion of celestial bodies. The chapter begins by introducing Newton's Law of Universal Gravitation and explaining the significance of the Gravitational Constant (G). It further explains the characteristics of gravitational forces, including the inverse square law and the interaction between multiple bodies. The chapter delves into Kepler's Laws of Planetary Motion and their derivation from the principle of Universal Gravitation. Special attention is given to the concept of orbital motion and escape velocity. Lastly, the chapter concludes by discussing the geosynchronous and heliocentric orbits, their applications, and the stability of these orbits over the long term.

REVIEW QUESTIONS

1. Explain Newton's Law of Universal Gravitation. How does it impact the motion of celestial bodies?

2. What is the significance of the Gravitational Constant (G) and how does it factor into calculations of gravitational forces?
3. Describe Kepler's Laws of Planetary Motion. How can these laws be derived from the principle of Universal Gravitation?
4. Define escape velocity. Why is it an important concept in the field of space exploration?
5. Discuss the differences between geosynchronous and heliocentric orbits. What are their respective applications and how does their stability affect their long-term usage?

REFERENCES

1. Abdel-Basset, M., Mohamed, R., Azeem, S. A. A., Jameel, M., & Abouhawwash, M. (2023). Kepler optimization algorithm: A new metaheuristic algorithm inspired by Kepler's laws of planetary motion. *Knowledge-Based Systems*, *268*, 110454.
2. Baez, A. V. (1960). Graphical derivation of the inverse-square law of gravitation from an elliptic orbit and Kepler's law of areas. *American Journal of Physics*, *28*(3), 254–259.
3. Barker, B. M., & O'connell, R. F. (1970). Derivation of the equations of motion of a gyroscope from the quantum theory of gravitation. *Physical Review D*, *2*(8), 1428.
4. Borowski, T. (2013). Actual picture of the law of universal gravitation and the quantum gravity theory describing the real state of the universe. *International Letters of Chemistry, Physics and Astronomy*, *11*.
5. Brunini, A. (2023). The effect of the giant planets on the dynamical evolution of the mutual orbits of trans-Neptunian binaries. *Monthly Notices of the Royal Astronomical Society*, *522*(2), 3067–3075.
6. Bruno, D., Hobbs, S. E., & Ottavianelli, G. (2006). Geosynchronous synthetic aperture radar: Concept design, properties and possible applications. *Acta Astronautica*, *59*(1–5), 149–156.
7. Carney, B. W., Latham, D. W., & Laird, J. B. (1988). A survey of proper-motion stars. V-Extreme-velocity stars and the local galactic escape velocity. *Astronomical Journal (ISSN 0004–6256), vol. 96, Aug. 1988, p. 560–566.*, *96*, 560–566.
8. Colombo, C. (2015). Long-term evolution of highly-elliptical orbits: luni-solar perturbation effects for stability and re-entry. In *Proceedings of the 25th AAS/AIAA Space Flight Mechanics Meeting* (pp. 3117–3140). Williamsburg: AAS-15–395.
9. Dobrovolskis, A. R., Alvarellos, J. L., & Lissauer, J. J. (2007). Lifetimes of small bodies in planetocentric (or heliocentric) orbits. *Icarus*, *188*(2), 481–505.
10. Hsiang, W. Y., & Straume, E. (2014). Revisiting the mathematical synthesis of the laws of Kepler and Galileo leading to Newton's law of universal gravitation. *arXiv preprint arXiv:1408.6758*.
11. Hsiang, W. Y., Chang, H. C., Yao, H., & Lee, P. S. (2015). Re-establishing Kepler's first two laws for planets in a concise way through the non-stationary Earth. *European Journal of Physics*, *36*(4), 045006.

12. Ikegaya, N., Ito, K., & Sandberg, M. (2022). Rigorous mathematical formulation of net escape velocity and net escape probability determining a macroscopic concentration. *Indoor air*, *32*(7), e13072.
13. Johnson, L. R., & Litehiser, J. J. (1972). A method for computing the gravitational attraction of three-dimensional bodies in a spherical or ellipsoidal Earth. *Journal of Geophysical Research*, *77*(35), 6999–7009.
14. Kempton, E. M. R., & Rauscher, E. (2012). Constraining high-speed winds in exoplanet atmospheres through observations of anomalous Doppler shifts during transit. *The Astrophysical Journal*, *751*(2), 117.
15. Kononogov, S. A., & Mel'nikov, V. N. (2005). The fundamental physical constants, the gravitational constant, and the SEE space experiment project. *Measurement Techniques*, *48*(6), 521–531.
16. Krittinatham, W., & Kaewkong, K. (2015). A Case Study of High School Students' Astrophysical-Conception Survey on the Kepler's Second Law of Motions and Newtonian Mechanics in Phayao. *Suranaree J. Sci. Technol*, *22*(2), 135–142.
17. Lecar, M., Franklin, F., & Murison, M. (1992). On predicting long-term orbital instability-A relation between the Lyapunov time and sudden orbital transitions. *Astronomical Journal (ISSN 0004–6256), vol. 104, no. 3, Sept. 1992, p. 1230–1236.*, *104*, 1230–1236.
18. Lim, E., Ito, K., & Sandberg, M. (2014). Performance evaluation of contaminant removal and air quality control for local ventilation systems using the ventilation index Net Escape Velocity. *Building and Environment*, *79*, 78–89.
19. Lissauer, J. J. (2007). Planets formed in habitable zones of M dwarf stars probably are deficient in volatiles. *The Astrophysical Journal*, *660*(2), L149.
20. Lombardi, A. M. (2018). Alexandre Koyré, Kepler's Reader Without Prejudices. Harmony of the World, Music of the Heavens. *Hypotheses and Perspectives in the History and Philosophy of Science: Homage to Alexandre Koyré 1892–1964*, 225–242.
21. Lopez, G. (2007). Constant of motion, Lagrangian and Hamiltonian of the gravitational attraction of two bodies with variable mass. *International Journal of Theoretical Physics*, *46*, 806–816.
22. Marcacci, F. (2022). Understanding, disseminating, and interpreting Kepler: the role of Giovanni Battista Riccioli and the law of orbits.

Understanding, Disseminating, and Interpreting Kepler: The Role of Giovanni Battista Riccioli and the Law of Orbits, 307–316.

23. McComas, D. J., Carrico, J. P., Hautamaki, B., Intelisano, M., Lebois, R., Loucks, M., ... & Tyler, R. (2011). A new class of long-term stable lunar resonance orbits: Space weather applications and the Interstellar Boundary Explorer. *Space Weather*, *9*(11).
24. Molland, A. G. (1975). Seventeenth Century-Der Musiktheoretiker Johannes Kepler. By Michael Dickreiter. Neue Heidelberger Studien zur Musikwissenschaft, Band 5. Berne and Munich: Francke Verlag, 1973. Pp. 252. 48 Swiss francs. *The British Journal for the History of Science*, *8*(1), 79–80.
25. Mošna, F. (2014). FROM NEWTON TO KEPLER. One simple derivation of Kepler's laws from Newton's ones. *Italian Journal of Pure and Applied Mathematics*, *32*, 393–400.
26. Nauenberg, M. (2005). Hooke's and Newton's contributions to the early development of orbital dynamics and the theory of universal gravitation. *Early Science and Medicine*, *10*(4), 518–528.
27. Papaxanthis, C., Pozzo, T., Popov, K. E., & McIntyre, J. (1998). Hand trajectories of vertical arm movements in one-G and zero-G environments Evidence for a central representation of gravitational force: Evidence for a central representation of gravitational force. *Experimental Brain Research*, *120*, 496–502.
28. Sannikova, T. N. (2021). Central field motion with perturbing acceleration varying by the inverse square law: application to the Yarkovsky effect. *Astronomy Reports*, *65*(4), 312–321.
29. Schmöle, J., Dragosits, M., Hepach, H., & Aspelmeyer, M. (2016). A micromechanical proof-of-principle experiment for measuring the gravitational force of milligram masses. *Classical and Quantum Gravity*, *33*(12), 125031.
30. Setyadin, A. H., Suryana, T. G. S., Utari, S., Efendi, R., Liliawati, W., & Utama, J. A. (2020). Identifying K-10 Students' Learning Difficulties on Learning Kepler's Law using Worksheet: Is It Worth?. In *Journal of Physics: Conference Series* (Vol. 1467, No. 1, p. 012051). IOP Publishing.
31. Shigemitsu, R., Tagawa, T., & Ozoe, H. (2003). Numerical computation for natural convection of air in a cubic enclosure under combination of magnetizing and gravitational forces. *Numerical Heat Transfer, Part A: Applications*, *43*(5), 449–463.

32. Spolter, P. (2011). Kepler's second law and conservation of angular momentum. *Physics Essays, 24*(2), 260–267.
33. Stern, S. A. (2005). Regarding the accretion of 2003 VB12 (Sedna) and like bodies in distant heliocentric orbits. *The Astronomical Journal, 129*(1), 526.
34. Such, J. (1990). The Idealizational Conception of Science and the Structure of the Law of Universal Gravitation. In *Idealization II: Forms and Applications* (pp. 125–130). Brill.
35. Tagawa, T., Ozoe, H., Inoue, K., Ito, M., Sassa, K., & Asai, S. (2001). Transient characteristics of convection and diffusion of oxygen gas in an open vertical cylinder under magnetizing and gravitational forces. *Chemical Engineering Science, 56*(14), 4217–4223.
36. Trilling, D. E., Benz, W., Guillot, T., Lunine, J. I., Hubbard, W. B., & Burrows, A. (1998). Orbital evolution and migration of giant planets: modeling extrasolar planets. *The Astrophysical Journal, 500*(1), 428.
37. Vatansever, S., Dirik, A. E., & Memon, N. (2022). The Effect of Inverse Square Law of Light on ENF in Videos Exposed by Rolling Shutter. *IEEE Transactions on Information Forensics and Security, 18*, 248–260.
38. Vezzoli, G. C. (2001). Physical Consequences of a Momenta-Transferring Particle Theory of Induced Gravity and New Measurements Indicating Variation from Inverse Square Law at Length Scale of. 1 mm: Statistical Time Properties of Gravitational Interaction and Analysis Thereof. *arXiv Preprint Physics/0104018.*
39. Wilson, C. A. (1970). From Kepler's laws, so-called, to universal gravitation: empirical factors. *Archive for History of Exact Sciences, 6*(2), 89–170.
40. Wu, X., Famaey, B., Gentile, G., Perets, H., & Zhao, H. (2008). Milky Way potentials in cold dark matter and Modified Newtonian Dynamics. Is the Large Magellanic Cloud on a bound orbit?. *Monthly Notices of the Royal Astronomical Society, 386*(4), 2199–2208.
41. Zeebe, R. E. (2015). Highly stable evolution of Earth's future orbit despite chaotic behavior of the Solar System. *The Astrophysical Journal, 811*(1), 9.
42. Zhao, H., Li, B., & Bienaymé, O. (2010). Modified Kepler's law, escape speed, and two-body problem in modified Newtonian dynamics-like theories. *Physical Review D, 82*(10), 103001.

43. Zhou, H., Pal, A., Mehta, A., Nakano, H., Modigliana, A., Arampatzis, T., & Howland, P. (2018). Reconfigurable phased array antenna consisting of high-gain high-tilt circularly polarized four-arm curl elements for near horizon scanning satellite applications. *IEEE Antennas and Wireless Propagation Letters*, *17*(12), 2324–2328.

INDEX